OPEN LIBRARY ⊚ SCIENCE

MOLECULES OF LIFE

OPEN LIBRARY ⊚ SCIENCE

General editor Frank Barnaby

MOLECULES OF LIFE

John Stares

Illustrations by Sandra Barnaby

GEOFFREY CHAPMAN

LONDON   1972

Geoffrey Chapman Publishers
35 Red Lion Square, London WC1 4SJ

Geoffrey Chapman (Ireland) Publishers
5-7 Main Street, Blackrock, County Dublin

ISBN 0 225 65855 0

First published 1972

I should like to express my gratitude to Dr Robert Cook who fostered my interest in biochemistry at Dundee University and who was a source of continual help and encouragement during my studies. I should also like to thank all the friends and colleagues, too numerous to mention individually, who made the writing of this book possible. Also, my thanks to Rita Lloyd who typed the manuscript.

John Stares

This book is set in 10 on 12pt, Times Roman and printed in Great Britain by A. Wheaton & Company, Exeter.

Contents

Suggestions for Further Reading

The number of books available on structural biochemistry is vast, and the following will provide a starting-point for further study of the subject. They all contain chapters or sections on the structural aspects, and their further reading lists will provide material for an even deeper study.

Outlines of Biochemistry by Eric E. Conn and P. K. Stumpf (John Wiley, 1963).

Introduction to Modern Biochemistry by P. Karlson (Academic Press, 1965).

The Chemistry of Life by Steven Rose (Pelican, 1970).

The Structure of Life by Royston Clowes (Pelican, 1967).

Chemical Background for the Biological Sciences by Emil H. White (Prentice Hall, 1970).

The Biological Role of the Nucleic Acids by David Cohen (Edward Arnold, 1965).

In addition to general books on biochemistry, many articles from scientific journals deal with specific aspects in more detail. Such articles from *Scientific American* are available as separate reprints, and the following should form an introduction to these:

The Chemical Structure of Proteins by W. H. Stein and S. Moore, *Scientific American*, February 1961.

The Nucleotide Sequence of a Nucleic Acid by R. W. Halley, *Scientific American*, February 1966.

The Genetic Code III by F. H. C. Crick, *Scientific American*, October 1966.

1
Some Basic Chemical Concepts

The modern science of chemistry has its foundations in the late eighteenth century when, with the final decline of the ancient doctrines of alchemy, scientists began to apply systematic observation and quantitative experimental methods to the study of the composition of the substances which made up the world around them. Today, chemistry can be defined as the science concerned with the structure and composition of the various kinds of matter and with the changes and transformation which matter undergoes.

Chemistry is an extremely wide field of study, embracing, at its extremes, the sciences of physics and biology. It can be divided into three main branches. Physical chemistry is concerned with the dependence of physical properties, such as energy and mass, on the composition and transformations of matter. Inorganic chemistry is the study of the actual properties and structures of matter and of the mechanisms of the changes and transformations. And organic chemistry is the science of the properties and reactions of the compounds of just one major element—carbon.

Organic chemistry is the branch of chemistry which merges with the science of biology, for almost all of the substances which make up the structures of living organisms are compounds of carbon. And in recent years a new science has arisen from the point at which chemistry and biology meet, a science in its own right, called biochemistry, the study of life and living organisms by chemical methods. Biochemists study all the chemicals, some quite simple but others vastly complex, which make up living organisms and the ways in which the compositions and reactions of these biochemicals are involved in the structures, functions and workings of systems which we recognize as being 'alive'.

The immense subject matter of biochemistry can be divided into two major areas. Structural biochemistry is concerned with the precise chemical structures of the many substances which make up living systems, while dynamic biochemistry involves the chemical changes, collectively called the metabolism, which occur in these systems. This book is devoted to the structural aspects of biochemistry, the composition and structures of biological compounds and the ways in which they are involved in the fabric of living organisms. But appreciation of structural biochemistry requires a working knowledge of some basic chemical principles, of the unique chemistry of carbon and of some elementary concepts of organic chemistry, and so the first two chapters will provide this essential background.

Elements and compounds

The composition of all matter in the universe is based on about a hundred substances called elements; these cannot be decomposed by chemical means into simpler substances and cannot be converted into each other. For example, the element oxygen cannot be split into other simpler substances, and cannot be converted into other elements such as hydrogen or lead. Chemists represent each element by a special symbol consisting of the initial letter of its name or its Latin or Greek name; thus oxygen is represented by O, hydrogen by H, carbon by C, lead by Pb and so on. Two or more elements joined together by a chemical reaction form a chemical compound, which can be split into its constituent elements only by another chemical reaction, and not by physical processes such as cutting. Compounds are represented by a combination of the chemical symbols for their elements, so that water, which contains two parts of hydrogen and one part of oxygen is written H_2O, and common table salt, sodium chloride, is written NaCl, Na being the chemical symbol for sodium and Cl the symbol for chlorine.

Atoms and molecules

The smallest possible amount of an element which can exist on its own and which can show the characteristic chemical properties of that element is called an atom. Today the atom can be further split into much smaller particles, but these no longer show the properties of the element. Similarly, the smallest amount of a compound which still retains its characteristic properties is called a molecule, so that molecules are made up of two or more atoms of elements. A molecule of water is made up of two atoms of hydrogen and one atom of oxygen, and a molecule of common salt consists of one atom each of sodium and chlorine. Sometimes two or more atoms of the same element combine to form a molecule of the element; molecules of oxygen and hydrogen for example each consist of two atoms of the element, and are thus written as O_2 and H_2.

Atoms are very small particles indeed; the hydrogen atom, for example, is only about one ten millionth of a millimetre in diameter. Such a small size is not readily comprehensible, but some idea of this size may be gained from the fact that just 56 grams of the element iron (a little less than two ounces) contain 602,000,000,000,000,000,000,000 atoms. If atoms were the size of garden peas, this number would be sufficient to blanket the entire surface of the earth to a depth of 10 feet, 60 times over.

Because even a microscopic piece of matter contains such large numbers of atoms, one might be forgiven for thinking that atoms are hard solid spheres, and that the matter they make up is a solid continuous substance, which was just the view held by the ancient 'scientists'. But in fact atoms are themselves made up of much smaller particles called protons, neutrons and electrons, and most of the atomic volume is just empty space.

The structure of the atom
The structure of the atom worried scientists for a very long time. When the existence of electrons as sub-atomic particles was demonstrated, scientists still stuck to their idea of a solid sphere, and put forward the notion that the atom was rather like a 'plum pudding'—a large mass of material with the electrons scattered in it like 'plums'. But later it was demonstrated that atoms contain a small, very dense region, much denser and with a much higher concentration of positive electric charge than could be explained on the basis of the 'plum pudding' model, and so a new theory was put forward, suggesting that the atom was like the solar system, with most of its mass concentrated in a nucleus at the centre, similar to the sun, and with the electrons moving in orbits round it like the planets. The fact that the orbits of the electrons ranged far out from the nucleus meant that most of the volume of the atom was in fact empty space, just like the solar system which has a few planets contained in a vast volume of space. Although this model is considerably modified today, it is sufficient for the purposes of biochemists to regard the atom as a

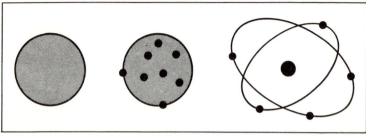

1. The early concept of the structure of the atom was of a solid sphere (*left*). When electrons were discovered as sub-atomic particles, they were thought to be simply embedded in the sphere, like a plum pudding (*centre*), but the modern concept is of a dense central nucleus with the electrons orbiting round it (*right*)

nucleus with orbiting electrons.

The nucleus of the atom is composed of small particles, the protons and neutrons. Both these particles have approximately the same mass, but while the proton carries a positive electric charge, the neutron is electrically neutral. Neutrons are usually present in the nucleus in slightly larger numbers than protons. The electron also carries an electric charge, negative with respect to the charge of the proton, but of the same magnitude. However, it has a mass of only 1/1836 of that of a proton. Atoms of different elements differ in that they contain different numbers of these sub-atomic particles, but in any given atom, the number of protons is always the same as the number of electrons, so that the whole atom is electrically neutral.

The smallest and simplest of all atoms in the universe is the hydrogen atom, consisting of just one proton in its nucleus and one electron orbiting it. There are no neutrons in the hydrogen atom. Helium, the second simplest atom, has a nucleus containing two protons and has two neutrons and two electrons orbiting it. Lithium, a metal element and the next in line, has three protons and four neutrons in its nucleus, and thus three orbiting electrons. The number of protons in the nucleus of any atom is called the atomic number, and as the atoms become larger and more complex, the atomic number increases until the heaviest elements contain about a hundred protons, together with the same number of electrons, and a slightly larger number of neutrons.

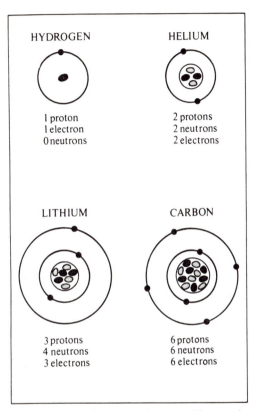

2. *Composition of some simple atoms.* The atomic number is the number of protons in the nucleus, and there is always an equal number of orbiting electrons. The outer shell of electrons is the valence shell which is involved in chemical bonding

The Periodic Table

By the middle of the nineteenth century it was noticed that there were certain groups of elements with similar properties. Lithium, sodium and potassium, for example, were all highly reactive soft metals, and chlorine, bromine and iodine all closely related non-metals. After considerable effort to find an explanation for these similarities it was found that by listing the elements in ascending order of their atomic numbers, and then arranging this list in rows of eight elements (starting with lithium, as hydrogen and helium did not fit into this scheme) and starting a new row with the ninth, the seventeenth and so on, a definite pattern emerged. This arrangement is the now familiar Periodic Table of the Elements, and all the elements beneath each other in a vertical row of the table, called a group, are seen to have similar properties. Lithium, sodium and potassium fall beneath each other in the same group, as do chlorine, bromine and iodine. Furthermore, along a horizontal row of the table, called a period, the properties of the elements change in a definite gradual manner. In one row for instance, sodium, magnesium and aluminium are all metals, silicon, phosphorus, sulphur and chlorine are all non-metals, and the last element in this row, argon, is a very unreactive gas. Although a number of elements, called the transition elements, did not fit into the original plan, and the table had to be modified to accommodate them, this scheme still forms the basis of the classification of the elements in modern chemistry.

Apart from showing similarities in properties between the elements, the Periodic Table can provide a great deal of information about the behaviour of the elements in chemical reactions. Because the elements are arranged in increasing order of atomic number, they are therefore arranged in increasing order of the number of electrons in their structures, as this number is the same as that of the protons in the nucleus. And it is the number of electrons in an atom which largely determines how it will react when brought into contact with another atom.

Electrons in an atom occupy certain fixed orbits or so-called shells around the nucleus, and each electron shell is able to hold a certain number of electrons. The first shell is capable of containing up to two electrons. If an atom consists of more than two electrons, the extra ones occupy the second shell, which can hold a maximum of eight electrons, and in atoms with more than 10 electrons, a third shell with a capacity of 10 electrons is involved, and so on. When considering how atoms react with each other, it is the outermost, partially filled, electron shell which is important; this shell is called the valence shell, and in the Periodic Table, all the elements in a particular vertical group have the same number of electrons in their valence shell. Lithium, sodium and potassium each have one electron in the outer shell of their atoms, and chlorine, bromine and iodine each

Group \ Period	I	II	transition elements	III	IV	V	VI	VII	VIII
1	1 **H** Hydrogen								2 **He** Helium
2	3 **Li** Lithium	4 **Be** Beryllium		5 **B** Boron	6 **C** Carbon	7 **N** Nitrogen	8 **O** Oxygen	9 **F** Fluorine	10 **Ne** Neon
3	11 **Na** Sodium	12 **Mg** Magnesium		13 **Al** Aluminium	14 **Si** Silicon	15 **P** Phosphorus	16 **S** Sulphur	17 **Cl** Chlorine	18 **Ar** Argon
4	19 **K** Potassium	20 **Ca** Calcium	transition elements	31 **Ga** Gallium	32 **Ge** Germanium	33 **As** Arsenic	34 **Se** Selenium	35 **Br** Bromine	36 **Kr** Krypton

3. *The periodic table of the elements.* Only the first four periods have been shown, as these contain nearly all the elements which are involved in biochemical compounds. Each element is identified by its name and chemical symbol, and the atomic number is given in each case. Elements with atomic numbers 21 to 30 inclusive have been omitted, as these form the first period of the transition elements

have seven. Along each horizontal row, the number of valence electrons increases in a regular way, just as the atomic number increases. The elements at the end of each horizontal row, forming the last vertical group, all have their valence shells filled to capacity, and it is significant that all of these elements are extremely stable, unreactive gases, popularly called the inert or noble gases. While most other atoms will react with each other when they come into contact, the atoms of these elements will not, and they are even loathe to react with each other to form molecules with two atoms as do hydrogen and oxygen.

Ionic and covalent bonds

When two atoms react together to form a molecule, there is an interaction between their electrons. Each atom aims to attain a state where its outer valence shell resembles that of a corresponding inert gas, as this electronic structure bestows great stability on the atom, much greater stability than it has with a partially filled valence shell. For example, a sodium atom has one electron in its valence shell, and if it can lose this electron, the valence shell will be empty, and its next inner shell will become the outermost one. This shell is filled to capacity with a similar electronic structure to that of the inert gas neon, which occupies a position in the Periodic Table at the end of the horizontal period before the one which contains sodium. Chlorine on the other hand has seven electrons in its outer shell, and if it can gain an electron from somewhere, it will have a completely filled outer valence shell resembling the valence shell of the rare gas argon which is at the end of the same horizontal period in which the chlorine atom occupies a place. So an atom of sodium and an atom of chlorine can 'join forces', the sodium giving an electron to the chlorine, so that both atoms benefit, and in fact when sodium and chlorine come into contact there is a violent reaction resulting in the formation of sodium chloride, or common salt.

What has happened is that the atoms have lost their electrical neutrality and have become electrically charged particles called ions. The sodium atom, having lost an electron, now has one more positively charged proton in its structure than it has negatively charged electrons, and so it becomes a positively charged ion or cation with one unit of positive charge. It is now written Na^+, to show that it carries a positive charge. Chlorine, having accepted an electron, now has a surplus of electrons over protons and has become a negatively charged chlorine ion or anion, written Cl^-. Because of the elementary law of physics which states that opposite charges attract each other, a sodium ion will be attracted to a chlorine ion, resulting in the formation of what is known as an ionic chemical bond between the two ions.

In a similar way, an atom of the metal calcium, which has two electrons in its outer valence shell, will give them both up, one to each of two atoms of chlorine, so that each chlorine anion will form an ionic bond with the calcium cation. Generally those elements on the right-hand side of the Periodic Table, with more than four electrons in their valence shell, will accept electrons and form negatively charged anions, while those on the left-hand side of the table, with fewer than four electrons in their outer shells, will try to lose their valence electrons and become positively charged cations.

The transfer of electrons from one atom to another involves the expenditure of a considerable amount of energy, the energy requirement being greater as more electrons are to be lost or gained. Thus it is very rare for more than three electrons to be lost or gained by an atom in the formation of an ionic bond. So there is another form of chemical bond in which, instead of transferring electrons, two atoms will share the same electrons in an attempt to attain an inert gas electronic structure in their outer valence shells. This form of chemical bond usually occurs between identical atoms, or between atoms of elements occupying positions in the top right-hand corner of the Periodic Table, and is called the covalent bond.

A covalent bond is formed by a pair of electrons, one of the pair being contributed by each atom involved in the bond, and the shared pair of electrons behave as though they are in the outer valence shells of both atoms at the same time. Thus two atoms of chlorine can each contribute one of their seven valence electrons to a covalent bond, such that each atom then has six valence electrons, plus the shared pair, in its outer shell, and this electronic structure is again similar to that of the outer shell of argon. The result of this electron sharing is a molecule of chlorine, Cl_2. Two hydrogen atoms can each contribute their single electrons to a covalent bond, so that they both appear to have two electrons in their valence shells, just as does helium, another of the rare gases, and in this way a hydrogen molecule is formed.

Carbon has an outer electron shell containing four electrons, and

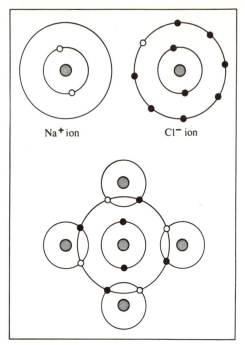

Na⁺ ion   Cl⁻ ion

4. *Ionic bond between sodium and chlorine.* An
electron has been lost by the sodium atom, so that
it now carries a net positive charge and is a sodium
ion, with a new outer electron shell resembling that
of the inert gas neon. The chlorine atom has
accepted an electron from sodium, giving it a net
negative charge, and the chlorine ion has an outer
electron shell resembling that of the inert gas argon.
Covalent bonds (*below*) are formed by the sharing
of a pair of electrons. The central carbon atom is
therefore able to form four covalent bonds with
hydrogen atoms, to form a molecule of methane

so the formation of ionic bonds involving the transfer of four electrons is very unlikely from energetic considerations. So carbon is
almost always involved in covalent bonding when it joins to other
atoms to form molecules. It can for example form covalent bonds
with four hydrogen atoms, each hydrogen contributing its single
electron to a bond, and the carbon contributing one of its valence
electrons to each of four bonds. Thus each hydrogen atom appears

to have two valence electrons, one of its own and one from the carbon, and the carbon appears to have an outer shell of eight electrons, four of its own and four from the hydrogen atoms, so that all the atoms now have electron structures in their valence shells similar to those of the corresponding rare gases. This compound, called methane and written CH_4, is the simplest of the organic compounds containing only the elements carbon and hydrogen, generally called the hydro-carbons.

The number of electrons which an atom can donate or accept in an ionic bond or can contribute to covalent bonds is called the valency of that atom. Thus chlorine can accept one electron from another atom to form a bond and is called monovalent with a valency of one. Similarly hydrogen can contribute one electron to a bond and is therefore also monovalent. Carbon has four electrons to contribute in covalent bonds and therefore has a valency of four—it is tetra-valent. Looked at another way, the valency of an atom is its combin-ing power with other atoms, in terms of hydrogen atoms or other equivalent monovalent atoms.

Atomic weights

Because elements are made up from atoms with different numbers of protons and neutrons in their nuclei, each different atom has a different weight. Each atom is therefore said to have a characteristic atomic weight. Although in some ways it would be sensible to use the actual weight of an atom as its atomic weight, these weights would be far too small to be used in practice. The hydrogen atom for example weighs only 1.67×10^{-24} grams and it would be impossible for a scientist to weight out such a small quantity. So a useful set of standard weights was devised in which the hydrogen atom was set as the standard with a weight of one unit, and the atomic weights of all the other elements were expressed as the number of times they were heavier than the hydrogen atom. On this scale, the atomic weight of the oxygen atom was 16, of carbon 12, of sodium 23, and so on.

Although this system still forms the basis of the calculation of atomic weights, the standard has been changed over the years, and the present day standard is taken as 1/12 of the weight of an atom of carbon instead of one atom of hydrogen. On this scale the atomic weight of hydrogen is 1.008 units (called atomic mass units, or *AMU*) and of sodium 22.9898 AMU. Molecular weights, the weights of molecules formed from two or more atoms bonded together, are simply calculated by adding together all the atomic weights of the atoms in the molecule. The molecular weight of the hydrogen molecule is thus 2.016—twice the weight of the hydrogen atom because there are two atoms in the molecule, and of methane 16.032, the sum of the atomic weights of one carbon atom and four hydrogen atoms.

Acids and bases

In later chapters in this book, which deal with the many complex and varied biochemicals which make up living organisms, we shall constantly come across the terms acid, base, and salt, and fundamental chemical reactions called oxidation and reduction reactions, so it will be useful to discuss these terms briefly at this stage. In the Middle Ages, alchemists recognized that there were two groups of compounds which could easily be classified in terms of their properties. One group comprised compounds which had a sour taste, dissolved metals, and were able to change the colour of solutions prepared from vegetable material (indicators). Members of this group were called acids. The other group, called the bases, had a brackish taste, only dissolved a few special metals, and when added to indicators changed them to a different colour from that produced by acids. These alchemists also found that bases react with acids to produce compounds which have neither acidic nor basic properties— no characteristic taste, no capacity to dissolve metals, and no ability to change the colour of indicators. These compounds are called salts.

This classification of substances into acids or bases is still with us, although we now have different ways of defining them. We define an acid as a substance which, in solution, can split, or dissociate, into ions, one of which is the hydrogen ion H^+. Because the hydrogen atom consists only of one proton and one electron, when it loses an electron to become a positively charged ion, all that is left is a proton, so we can call a hydrogen ion simply a proton. Acids can now be redefined as substances which dissociate in solution, giving rise to protons. An example of an acid is hydrogen chloride, which, when dissolved in water, dissociates into a chloride ion, Cl^-, and a proton.

For the purposes of this book, the best definition of a base is a substance which will accept a proton from some external source.

Thus ammonia, a compound of nitrogen and hydrogen written NH_3, is a base because in solution in water it can accept a proton and become an ammonium ion NH_4^+. Bases used to be defined as substances which in solution, gave rise to hydroxyl ions, OH^-, and potassium hydroxide was thus classified as a base because it dissociated into a potassium ion K^+, and a hydroxyl ion. But using our definition, potassium hydroxide is a base because the hydroxyl ions which it gives rise to can accept protons to form water.

When a base reacts with an acid it is said to neutralize it, with the formation of a salt and water. Thus when hydrogen chloride in solution reacts with potassium hydroxide in solution, the proton and the hydroxyl ions react together to form water, and the potassium ions and the chloride ions are left in the solution, so that the solution is said to contain the salt potassium chloride. If the water is distilled off from the solution, the potassium and the chloride ions come together in ionic bonds, because they carry opposite electric charges, and form molecules of potassium chloride.

In fact all ionic compounds, when dissolved in water, dissociate into their ions, and such compounds can be called electrolytes. If two wires are immersed into a solution of an electrolyte, and a current passed through the wires from a battery, it is found that the positively charged ions tend to migrate through the solution towards the negatively charged wire (called the cathode), and the negatively charged ions move towards the positively charged wire, the anode. Most salts and some acids, such as hydrogen chloride, which dissociate almost completely into their ions when dissolved in water, are said to be strong electrolytes, while other substances which only partially dissociate—acetic acid is such a compound—are called weak electrolytes.

The acidity of a solution is a very important measurement in both chemistry and in biochemistry. It is measured as the concentration of hydrogen ions in a solution, and is expressed on a scale called the pH scale. The pH value of a solution is calculated by measuring the hydrogen ion concentration, and then calculating the minus common logarithm of this value (usually by reference to mathematical tables rather than calculation) and taking this figure as the pH. Solutions which are neutral, that is are neither acidic nor basic, are found to have a pH value of seven. Strong acids have a much higher concentration of hydrogen ions, so that their pH values will be much less

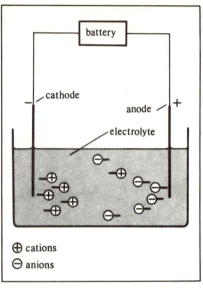

5. When an electric current is passed through a solution of an electrolyte, the ions migrate towards the electrodes, positively charged cations collecting at the negative cathode, and negatively charged anions collecting at the positive anode. This principle is used in silver plating, the object to be plated being made the cathode, and positively charged silver ions from a solution of a silver salt collecting on it

than seven; in fact strong acids have a pH of about one. Strong bases, however, have a much lower hydrogen ion concentration, and hence pH values higher than seven, usually about thirteen. Weak acids and bases which do not dissociate so completely have pH values of perhaps six and eight respectively and are thus only slightly acidic or basic. Pure water and solutions of most true salts are neutral, with pH values of seven.

Oxidation and reduction

Oxidation and reduction reactions are encountered in all aspects of chemistry and biochemistry. There are a number of definitions of both types of reaction; oxidation can be defined as the addition of oxygen to a substance or the removal of hydrogen from it, and, conversely, reduction can be defined as the removal of oxygen or the addition of hydrogen. But the general definition is that the oxidation reaction involves the removal of an electron from an atom or ion, and the reduction reaction the addition of an electron. So when a chlorine atom gains an electron to become a chloride ion, it has undergone a reduction reaction, and when a sodium atom loses an electron to become a sodium ion, it has been oxidized. In all cases, oxidation reactions must be accompanied by reduction reactions, so that, in this example, both the sodium and the chlorine atoms must form ions when they come into contact with each other, and it would be clearly impossible for, say, only the sodium atoms to ionize.

Now that we understand the very basis of chemistry—the structure of atoms and the ways they can be joined together to form molecules —we can begin to look at the element carbon, which forms the basic of organic chemistry, and at how this element gives rise to such a diversity of organic compounds.

2
The Chemistry of Carbon

Structural biochemistry has its roots in the middle of the eighteenth century, when scientists were trying to discover the composition and function of living organisms. But because of the complexity of the problem, a deep appreciation of the compounds and the chemical reactions associated with living materials had to wait for the development of chemical theory and of specialist research techniques.

The initial impetus for the development of structural biochemistry came from a Swedish chemist Karl Scheele, who in the mid-eighteenth century isolated a number of natural substances from the tissues of both animals and plants. At the beginning of the nineteenth century, quantitative methods for the analysis of the elements in material were developed in the laboratories of Jöns Berzelius and Justus von Liebig, and Scheele's natural substances were all shown to contain carbon. With this discovery, attempts were made to synthesize these compounds in the laboratory from simple elements and compounds.

But it was believed at that time that the synthesis of natural substances from simple materials in a test tube was impossible, that these substances could only be made in living organisms through the agency of some 'vital force'. This doctrine was called vitalism, and the substances which could only be made by living systems were described as being 'organic'. The theory of vitalism was eventually disproved in 1828 when Friedrich Wöhler succeeded in synthesizing urea, a chemical present in urine of mammals, from simple inorganic chemicals. In fact this synthesis was quite accidental. Wöhler was trying to prepare ammonium cyanates by chemically combining potassium cyanate and ammonium sulphate, but instead found that he had prepared urea, thus showing that organic chemicals differed from inorganic ones only in complexity, that they owe their special properties to the unique nature of carbon and not to any mysterious force of nature. Wöhler's breakthrough was followed by the synthesis of acetic acid by Adolf Kolbe in 1844, and by the synthesis of several other organic chemicals by Marcellin Berthelot in the 1850s. Vitalism was dead, but the term 'organic chemistry' still remains to describe the chemistry of the compounds of carbon. And the term biochemistry came to refer to the chemical nature of living organisms.

Chemistry of carbon

As carbon is the major constituent of the biologically important compounds, its chemistry is of supreme importance in determining the properties of biochemicals. The unique behaviour of carbon in forming such a vast number of compounds—more compounds of carbon are known today than of all the other elements put together—is due not to its abundance in the earth's crust, but to its ability to form chemical bonds with itself, apparently without limit, forming very long chain-like molecules containing many tens or hundreds of atoms. These chains of carbon atoms are extremely stable and not easily broken down. The element silicon is some hundred and fifty times as abundant as carbon in the earth's crust, but although it will form chemical bonds with itself, the resulting molecules are much less stable, and the chains of silicon atoms cannot attain lengths of more than a few atoms; so the number of compounds of silicon is only a very small fraction of that of carbon compounds.

Another property of carbon which increases the possible number of compounds is that it can combine with almost equal facility with other elements such as hydrogen, oxygen and nitrogen, forming compounds which are also very stable. Silicon compounds are, again, not so stable and often cannot even exist as single discrete molecules; silicon dioxide, for example, tends to form large clumsy aggregates of molecules in an effort to gain stability, while the corresponding compound of carbon, carbon dioxide, is unusually stable as a single molecule.

From its position in the Periodic Table (see p. 13) it can be seen that carbon has an atomic number of six, which means that the carbon atom contains six electrons. Two of these completely fill the first electron shell, while the other four occupy and partially fill the outer valence shell and are therefore involved in chemical bonds with other atoms. The carbon atom is therefore tetravalent, and is able to form four covalent bonds with hydrogen atoms or their equivalent.

Because covalent bonds result from the sharing of a pair of electrons, they always have a definite spacial orientation about the central atom. In the case of carbon, the four bonds are so arranged that they point to the four corners of a regular tetrahedron, with the carbon atom in the centre. A simple model of this arrangement can be constructed by sticking four cocktail sticks into an orange, such that whichever way the orange is turned, it rests on a tripod of three sticks, with the points of the sticks all equal distances from each other, and then sticking four other oranges on the ends of the sticks. The central orange then represents the carbon atom, the sticks the covalent bonds and the other four oranges the other atoms involved in the chemical bonds. Throughout this book, chemical bonds will be represented, in diagrams of molecules, as short dashes linking the two atoms involved in the bond, and the atoms themselves will be represented by their chemical symbols.

Aliphatic hydrocarbons

If the four outer oranges in the above model represent hydrogen atoms, then the whole model represents a molecule of methane, a gas which is the principal constituent of marsh gas, written CH_4. Methane is the simplest of a very large series of organic compounds called the hydrocarbons, which, as the name suggests, contain only the elements carbon and hydrogen. If one of the hydrogen atoms in methane is replaced by another carbon atom, itself with three more hydrogen atoms attached to it, the molecule so formed is the next member of the series, another gas called ethane. Ethane can be written as CH_3CH_3; for convenience the dashes representing the covalent bonds between the carbon atoms and the hydrogen atoms can be omitted, as it is quite clear from this representation, called the chemical formula of the compound, that the carbon is bonded four times, three times to hydrogen atoms and once to the other carbon, and it is understood by writing formulae in this way that each of the covalent bonds has a tetrahedral orientation about the carbon atom. Hydrocarbons consisting of carbon atoms joined together in long chains are called aliphatic hydrocarbons.

The next hydrocarbon in the series contains three carbon atoms joined together in a chain, as though one of the hydrogens in ethane is replaced by a carbon atom with three hydrogens attached to it. Each carbon atom at the end of the chain has three hydrogen atoms

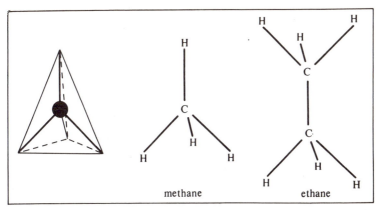

methane ethane

6. The four covalent bonds of the carbon atom point to the four corners of a
regular tetrahedron (*left*). If each of the bonds involves a hydrogen atom, the
molecule is methane; if two carbon atoms, each with their tetrahedral shape, are
joined together, the molecule is ethane. More and more carbon atoms can be
joined together in this way to make longer carbon chains, and if hydrogen is the
only other element present, all the molecules are hydrocarbons

bound to it, but the carbon atom in the middle is already involved in
two covalent bonds to the other two carbon atoms, and therefore
only has room for two hydrogen atoms. So this third hydrocarbon,
which is called propane, can be written $CH_3CH_2CH_3$. Similarly, the
fourth member, butane, can be written $CH_3CH_2CH_2CH_3$.

It will be noticed that a pattern begins to emerge from these
compounds; the number of hydrogen atoms in each compound is
always twice the number of carbon atoms, plus two, and we can
write a general formula C_nH_{2n+2}, where n represents the number of
carbon atoms in the chain. Thus if a hydrocarbon contains six carbon
atoms in a chain, it will contain $[(2 \times 6) + 2]$ hydrogen atoms, that is
14 hydrogens; it can thus be written as $CH_3CH_2CH_2CH_2CH_2CH_3$.

Such a series of compounds which can be represented by a general
formula, and where each member of the series differs from the
previous member by containing one more carbon atom and a
corresponding number of extra hydrogen atoms, is called a homologous series, and its members are called homologues. This homologous
series, where the molecules contain the greatest possible number of
hydrogen atoms for a given number of carbon atoms, is called

generally the alkanes, and its homologues are said to be saturated with hydrogen.

Grouping compounds into homologous series greatly simplifies the organization of organic chemistry, and we shall meet other such series later on. Another interesting point about homologous series is that the chemical and physical properties of each member are slightly different from those of the previous member, and that properties change progressively as the homologues become larger. In the series just discussed the first few members are gases. Butane, the hydrocarbon with four carbon atoms, is a liquid with a fairly low boiling point, and as the carbon chain grows longer, the boiling point of the compounds rises, until, when the carbon chain has reached a certain length, the compounds are solids, and as the chain grows even longer, the melting point of the solid hydrocarbons increases.

In all the compounds of carbon which have been described so far, each carbon atom is joined to four separate atoms, either hydrogens or other carbons, and so only one pair of electrons has been involved in each bond. A slightly different situation occurs however, when carbon combines with an atom such as oxygen. Oxygen, as can be seen from the Periodic Table (see p. 13), has an outer valence shell of six electrons, and hence has a valency of two—one atom of oxygen can combine with two atoms of hydrogen. A carbon atom can therefore combine with only two atoms of divalent oxygen instead of four atoms of monovalent hydrogen, and so each bond between carbon and oxygen must involve the sharing of two pairs of electrons, two electrons from the carbon and two from the oxygen. The bond between these two atoms is therefore a double covalent bond, and one atom of carbon can combine with two atoms of oxygen, forming two double bonds, and producing a molecule of carbon dioxide.

Yet another situation arises when carbon combines with an atom such as nitrogen, for nitrogen has a valence shell containing five electrons, three less than the corresponding rare gas configuration, and hence has a valency of three. The chemical bond between an atom of carbon and an atom of nitrogen is therefore a triple bond, involving three pairs of electrons, three from the carbon and three from the nitrogen. The carbon atom of course still has one electron free in its valence shell with which to combine in a single covalent bond with another monovalent atom.

As might be expected, molecules involving double and triple bonds do not have a tetrahedral shape. Using oranges and cocktail sticks again, a simple model of the carbon dioxide molecule can be constructed by sticking two sticks into each side of an orange, and then sticking another orange on to the ends of each pair of sticks, the outer oranges representing the oxygen atoms. It will be noticed that this molecule is a straight, flat, two-dimensional structure, quite different from the three-dimensional structure of the methane molecule.

Carbon atoms can themselves form double and triple covalent bonds with other carbon atoms by the same method of sharing two or three pairs of electrons between them. If two carbon atoms are joined by a double bond, each atom then has two spare valence electrons with which to form covalent bonds with hydrogen atoms, and this molecule, called ethene, is the simplest member of another homologous series of hydrocarbons called the alkenes. Ethene can be written $CH_2=CH_2$, the double dashes between the carbon atoms representing the double bond. Long chains of carbon atoms can be built up in this way, and can have one or more double bonds in the chain. The presence of each double bond means that the whole molecule contains less hydrogen atoms than a chain of carbon atoms all joined by single bonds, and so these compounds are said to be unsaturated with hydrogen, that is, the number of hydrogen atoms present is not the maximum possible for a given number of carbon atoms.

Carbon atoms joined by triple covalent bonds form the basis of yet another homologous series of hydrocarbons. The simplest member consists of two carbon atoms joined with a double bond, and hence containing only two hydrogen atoms, as each carbon atom has only one electron left in its valence shell to form the fourth bond. This molecule is called ethyne, usually referred to as acetylene, its older name, and the whole series is called the alkynes. They are of course all unsaturated, and to a greater extent than the alkenes.

So far only hydrocarbons consisting of straight chains of carbon atoms have been discussed, but carbon atoms can join together in other ways. For example, four carbon atoms can join in a straight chain of four atoms, as we have already seen, or they can join as a chain of three atoms, the fourth joining on to the middle carbon atom to form a branch. Hydrocarbons of this latter type are called, naturally, branched chain hydrocarbons. If the number of hydrogen

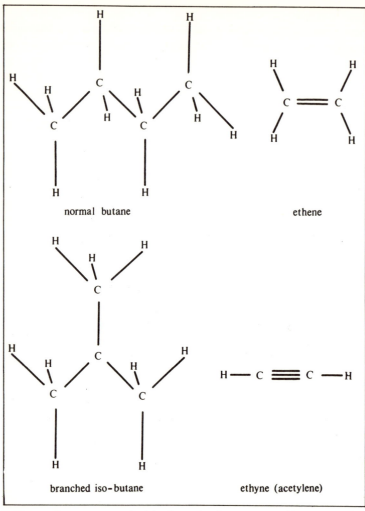

normal butane

ethene

branched iso-butane

ethyne (acetylene)

7. Four carbon atoms can be joined together in two ways, either as a straight chain (*top left*) or in a branched chain (*bottom left*). Although both these molecules have the same number of atoms, they are different compounds with different properties, and as the number of carbon atoms increases, there is more scope for branching. Carbon atoms can be joined by double (*top right*) or triple (*bottom right*) bonds, and long carbon chains can contain one or more of these bonds, forming unsaturated hydrocarbons

atoms in a branched chain molecule is counted, it will be found to be exactly the same as in the corresponding straight chain molecule with the same number of carbon atoms; in this example three of the carbons are joined to just one other carbon, and so can bond three hydrogen atoms each, while the carbon in the middle of the chain forming the branch, is joined to three other carbons, and so has room for only one hydrogen, making a total of 10 hydrogens, the same as in the straight chain butane. As the number of carbon atoms in a hydrocarbon increases, there is much more scope for branching, but in all cases the general formulae of the straight and the branched chain saturated hydrocarbons is C_nH_{2n+2}.

Straight and branched chain hydrocarbons, although consisting of the same number of carbon and hydrogen atoms, clearly have different structures, and so are different compounds with different chemical and physical properties. Compounds such as these, with the same number of atoms in their molecules, are called structural isomers, so that butane exists in two isomers. We can write the branched chain isomer as $CH_3CH(CH_3)CH_3$, the bracket indicating that a branch exists in the molecule, and all of the other hydrocarbons with branched chains can be written in a similar way. A molecule $CH_3CH(CH_2CH_3)$—CH_3 contains a branch of two carbon atoms. Formulae written in this way are called structural formulae, as they show the structure of the molecule, while a formula written simply as C_4H_{10} for butane is called the empirical formula, and shows only the number of atoms involved.

Cyclic and aromatic hydrocarbons

Carbon atoms can also be joined together so that, instead of forming straight or branched chains, they form rings. The simplest ring structures consist of carbon atoms joined together by single bonds, and can be thought of as derivatives of the alkanes. These molecules can be represented by the general empirical formula C_nH_{2n}; they will have two less hydrogen atoms than the corresponding straight chain molecule, because each carbon atom is joined to two carbon atoms in the ring and can thus only bond two hydrogen atoms. But because this is the maximum possible number of hydrogens in this structure, all of these cycloalkanes, as they are called, are saturated hydrocarbons. Sometimes two or more of these carbon rings join together forming bicycloalkanes and tricycloalkanes.

Much more important in organic chemistry are the ring structures related to a compound called benzene. These compounds are called the aromatic hydrocarbons, because when they were first discovered, it was observed that many of them had characteristic fragrant smells.

Benzene is the simplest of the aromatic hydrocarbons, consisting of

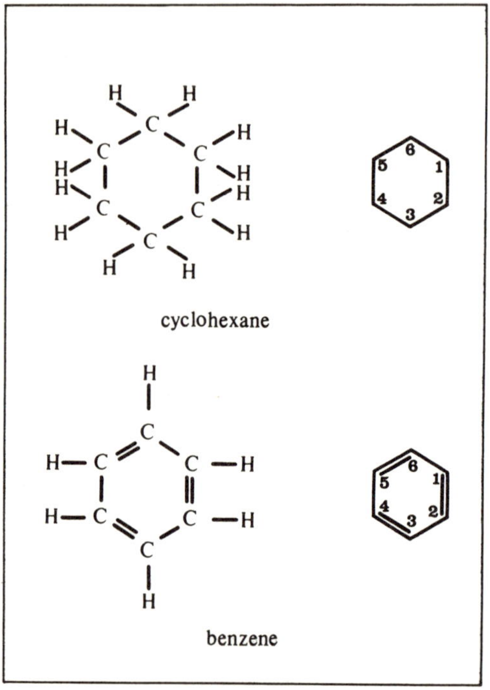

cyclohexane

benzene

8. Carbon atoms can also be joined together in rings. Cyclohexane is a saturated compound, while benzene is unsaturated, and forms the basis of the aromatic series of hydrocarbons. For convenience these ring structures are usually written as on the right, without the carbon atoms being shown

a six-membered ring of carbon atoms. It is usually written as a flat, regular hexagon, with the carbon atoms at the corners and single and double bonds alternating round the ring. Each carbon atom is therefore joined to another carbon atom on one side by a single bond and on the other side by a double bond, and hence has one valence electron left to form a covalent bond with hydrogen. The empirical formula of benzene is thus C_6H_6, and it is an unsaturated compound. This structure for benzene was postulated in 1865 by a German chemist August Kekulé, and it accounted for all the physical and chemical properties of the compound known at that time. Today, however, it is recognized that this representation is inadequate, that in fact each carbon atom is joined to each other carbon atom by a bond which is a hybrid between a single and a double bond, so that all of the six bonds in the structure are equivalent. (One way to picture this, although not quite correct, is to imagine each carbon atom joined to the next by $1\frac{1}{2}$ covalent bonds.) For the purpose of this book, the traditional representation of the benzene molecule is sufficient.

Aromatic ring structures can join together, just as can the cycloalkanes, and so there are further series of aromatic compounds based on bicyclic and tricyclic structures and so on. The benzene structure, together with examples of straight and branched chain hydrocarbons and cycloalkane, are shown in diagram 8 on page 30.

Functional groups

The hydrocarbons account for a vast number of organic compounds, and are to be found in most groups of biochemicals in one form or another. But the range of organic compounds and biochemicals is increased still further by the involvement of other elements. The most common of these other elements are oxygen and nitrogen, and the most convenient way of studying the role of these elements in organic chemistry and biochemistry is to think of these organized into simple structures called functional groups. Functional groups are convenient because the majority of the biochemicals which will be encountered in later chapters are based on a hydrocarbon structure with one or more of the functional groups substituted in place of the hydrogen atoms. It must be stressed, however, that the complex chemicals are not formed in practice simply by removing hydrogen atoms and joining on functional groups; methods of formation and preparation are usually much more complex, and the concept of functional groups is only a convenience for classifying and simplifying the study of the structural aspects of biochemistry. In the following section, and in diagrams later in the book, the letter R will be used to signify a hydrocarbon constituent or some other group in a molecule, where this constituent has no significance in the particular topic being discussed.

Organic compounds which contain the hydroxyl group, $-OH$, are called alcohols, after the most well-known member of this group, ordinary or ethyl alcohol found in beer, wine and spirits. Alcohols then can be thought of as hydrocarbons in which one of the hydrogen atoms has been replaced by the hydroxyl group, and three different types are recognized depending on the position of the hydroxyl group on the carbon chain. Primary alcohols have the general formula RCH_2-OH; the hydroxyl group is joined to the terminal carbon atom of a chain. Secondary alcohols have the formula $RCHR'-OH$, and in this case the hydroxyl group replaces the hydrogen on one of the carbon atoms in a straight chain. The third group, the tertiary alcohols have the formula $RC(R')R''-OH$, and here the $-OH$ group is joined to the carbon atom which forms the junction of carbon chains in a branched hydrocarbon. These different types of structures are set out more fully in diagram 9 on page 34.

Alcohols with only short carbon chains are all liquids soluble in water, while those with longer carbon chains are solids at room temperature and are insoluble or only very slightly soluble in water. This property again reflects a very important property of homologous series—the properties of the homologues change progressively as the molecule becomes larger. This solubility also shows that in general the hydrocarbons longer than a certain length are insoluble, and this property will be encountered a great deal in discussion of some of the complex biochemicals. Alcohols can also be classified according to the number of hydroxyl group involved in the molecule. These different compounds are called monohydric, dihydric and trihydric alcohols according to whether they contain one, two or three hydroxyl groups respectively.

As well as being a functional group joined to straight and branched chain hydrocarbons, the hydroxyl group can also replace a hydrogen atom in the benzene molecule, forming a compound called phenol. Phenol, and other molecules related to it, can ionize slightly, forming a negatively charged ion and a proton—hence these compounds are weak acids, and in fact an old name for phenol was carbolic acid.

A functional group consisting of a carbon atom joined by a double bond to an oxygen atom is called the carbonyl group, $>C=O$. The carbon atom still has two valence electrons available to form bonds with other atoms or groups of atoms, and depending on what is joined to this carbon, there are two classes of compounds containing the carbonyl group. If the carbon is joined to two other carbon atoms or chains of carbon atoms, that is if the carbonyl group is joined to the middle of a carbon chain, then the compounds are called ketones, while if the carbonyl group is at the end of a carbon chain, so that it is joined to a carbon atom and to a hydrogen atom, then the compounds are called aldehydes.

If a carbon atom is joined by a double bond to an oxygen atom, and by a single bond to a hydroxyl group, the whole functional group is called a carboxylic acid group, and this functional group forms the basis of a very important group of compounds, the carboxylic, or simply organic, acids. The presence of the carbonyl group and the hydroxyl group together on the same carbon atom bestows properties on the whole compound which are quite different from those with the separate groups. Each group modifies the properties of the other group; the carbonyl group does not react as it does in aldehydes or ketones, and the hydroxyl group ionizes to a much greater extent than it does in the alcohols, or the phenols. The hydrogen atom of the hydroxyl group can be lost as a proton, leaving the carboxylic group with a negative charge, so that the carboxylic acids are weak acids.

Just as there are different classes of the alcohols depending on how many hydroxyl groups are involved in the molecule, different carboxylic acids are recognized with different numbers of carboxyl groups. Thus there are dicarboxylic acids and tricarboxylic acids containing two and three carboxyl groups respectively.

Nitrogen is an essential element in many biochemicals, and the most important functional group containing nitrogen is the amine group $-NH_2$, where the atoms of hydrogen are joined by single bonds to a nitrogen atom. Primary, secondary and tertiary amines are recognized, but these designations are derived in a slightly different manner to the names of primary, secondary and tertiary alcohols. A secondary amine has a carbon atom or chain attached to the nitrogen atom in place of one of the hydrogen atoms, so that it can be written $R-NHR'$, the R' group being the same or different from the R group. Similarly a tertiary amine has another carbon atom or chain replacing the second hydrogen atom in the amine group, and so can be written $R-NR'R''$. Organic amines of these

FUNCTIONAL GROUP	COMPOUNDS		
hydroxyl — OH	R—$\overset{\overset{\displaystyle H}{\vert}}{\underset{\underset{\displaystyle H}{\vert}}{C}}$—OH primary alcohol	R—$\overset{\overset{\displaystyle H}{\vert}}{\underset{\underset{\displaystyle R'}{\vert}}{C}}$—OH secondary alcohol	R—$\overset{\overset{\displaystyle R''}{\vert}}{\underset{\underset{\displaystyle R'}{\vert}}{C}}$—OH tertiary alcohol
carbonyl $\overset{\diagdown}{\underset{\diagup}{C}} = O$	$\overset{\displaystyle R}{\underset{\displaystyle H}{\diagup\!\!\!\diagdown}} C = O$ aldehyde		$\overset{\displaystyle R}{\underset{\displaystyle R'}{\diagup\!\!\!\diagdown}} C = O$ ketone
carboxyl $-C\overset{\diagup\!\!\!O}{\diagdown_{OH}}$	$R-C\overset{\diagup\!\!\!O}{\diagdown_{OH}}$ monocarboxylic acid		$R-C\overset{\diagup\!\!\!O}{\diagdown_{O^-}}$ carboxylate anion
amine $-\overset{\overset{\displaystyle H}{\diagup}}{\underset{\underset{\displaystyle H}{\diagdown}}{N}}$	$R-\overset{\overset{\displaystyle H}{\diagup}}{\underset{\underset{\displaystyle H}{\diagdown}}{N}}$ primary amine	$R-\overset{\overset{\displaystyle H}{\diagup}}{\underset{\underset{\displaystyle R'}{\diagdown}}{N}}$ secondary amine	$R-\overset{\overset{\displaystyle R''}{\diagup}}{\underset{\underset{\displaystyle R'}{\diagdown}}{N}}$ tertiary amine $R-\overset{\overset{\displaystyle H}{\diagup}}{\underset{\underset{\displaystyle H}{\diagdown}}{N^+}}-H$ amine cation

9. *Common functional groups encountered in biochemistry.* On the right side of the diagram some of the ways in which these groups are involved in molecules are shown. The letters R, R' and R" represent the rest of the molecule, usually just a hydrocarbon chain

series are organic bases and so are able to accept a proton and become positively charged species $R-NH_3^+$. This property is one of the most important aspects of the role of the amines in biochemistry.

Chemical reactions

Some of the chemical reactions in which these functional groups can take part will be met in later discussion on the major groups of biochemicals, and so a brief survey will be given here. The number of reactions involved in biochemistry is very small, much smaller than the number of reactions available to and forming the field of study of the pure organic chemist.

The hydroxyl group can undergo an oxidation reaction (oxidation here can, for convenience, be thought of as the removal of hydrogen rather than the more general electron addition, although of course both definitions are equivalent) to form a carbonyl group. Oxidation of a primary alcohol will produce an aldehyde, because the new carbonyl group will have a hydrogen attached to it, while oxidation of a secondary alcohol will give rise to a ketone, with two carbon chains attached to the carbonyl carbon atom. Similarly, a carbonyl group can be reduced (by the addition of hydrogen) to a hydroxyl group, and if a carbonyl group is oxidized, it will produce a carboxyl group. Thus a series of oxidation reactions could go: alcohol to aldehyde to carboxylic acid, and reduction would be the reverse of this series.

Another important reaction of the carbonyl group is that with alcohols, or rather with the hydroxyl group of alcohols. These two groups can react to produce a compound called a hemiacetal, which can react with another hydroxyl group to produce the full acetal, or simply the acetal, and this important reaction sequence is shown in diagram 10.

The chemical reaction between an alcohol and a carboxylic acid is in many ways similar to the reaction between an inorganic acid and a base. In organic chemistry the 'salt' formed is called an ester, and can be written $R-CO.O-R'$, where R is the carbon chain of the carboxylic acid and R' the carbon chain of the alcohol. It is important to note that the carbon chain from the alcohol is joined to the oxygen atom which originally formed the hydroxyl group of the carboxylic acid, and that therefore the molecule of water which is formed during the formation of the ester is derived from the proton

10. *Acetal formation.* If a carbonyl group reacts with an alcohol (hydroxyl group), a hemiacetal is formed, and this compound can react with another hydroxyl group to yield a full acetal, or, simply, acetal

of the carboxylic acid and the whole hydroxyl group of the alcohol. The full point is written in the formula of the ester above to indicate that both of the oxygen atoms are in fact attached to the carbon atom of the ester linkage group. Aromatic alcohols, the compounds related to phenol, can also form esters with carboxylic acids, so that although they are able to ionize to form a proton, they are more closely related to the alcohols in their chemical properties than to acids.

Another salt which is of great importance in biochemistry is the ester of an alcohol and inorganic phosphoric acid, because most of the phosphorus which is formed in biologically important materials is in the form of phosphate salts. The mechanism of formation of these esters is the same as in the carboxylic esters; water is formed from the hydroxyl group of the alcohol and a hydrogen ion from the phosphoric acid, and the resulting salt is called an alcohol phosphate.

The final group of structures which contain elements other than carbon and hydrogen cannot strictly be said to contain a functional group. They are ring structures in which one or more of the carbon atoms in the ring have been replaced by atoms of oxygen, nitrogen or sulphur, and these structures which are more important in biochemistry than benzene compounds, are called the heterocyclic ring compounds. Ring systems of this type have quite different properties from analogous rings containing only carbon atoms, and they will be encountered a great deal in later chapters on the major groups of biochemicals.

Isomerism

As well as the great diversity of hydrocarbon compounds and organic compounds containing one or more of the functional groups just described, the range of organic molecules is increased even further by another factor—isomerism. Isomers are compounds which have the same empirical formulae, but different structural forms and hence different chemical and physical properties, and the straight and branched isomeric forms of hydrocarbons have already been mentioned as being examples of structural isomerism. In this final section of this chapter, two other forms of isomerism will be introduced. These are geometric and optical isomerism, which together can be called forms of stereoisomerism. Stereosomers have the same empirical formulae and structures, except that they have different arrangements of their atoms in space, and a few examples will be given here. Others will be met later on.

To understand geometric isomerism, it is necessary to imagine a plane drawn through a molecule and to examine which functional groups lie on either side of the plane. For example, if two carbon atoms are joined by a double covalent bond, a plane can be drawn through the two carbon atoms and through the 'middle' of the double bond. If we now imagine a hydrogen atom and a carboxylic functional group attached to each carbon atom, there must be one atom or group attached to each carbon atom on either side of the plane through the molecule. So two possibilities exist. If the carboxylic group of one carbon atom is on the opposite side of the plane to the carboxylic group of the other carbon atom, the carboxylic groups are said to be in the 'trans' configuration, and the molecule is a discarbocylic acid called fumaric acid. If on the other hand, both of the carboxylic groups are on the same side of the plane through the molecule, they are in the 'cis' configuration, and the molecule is another decarboxylic acid called maleic acid. Fumaric and maleic acid are then two geometric isomers, and this type of isomerism is often called cis-trans isomerism.

To appreciate optical isomerism, we must go back to the structure of a carbon atom and its covalent bonds forming a tetrahedron. In a molecule of methane, all of the four hydrogen atoms are equal and equivalent, and it is impossible to distinguish between them. If one of these hydrogens is replaced by, say, a chlorine atom, a molecule of methyl chloride is formed, and whichever hydrogen atom is

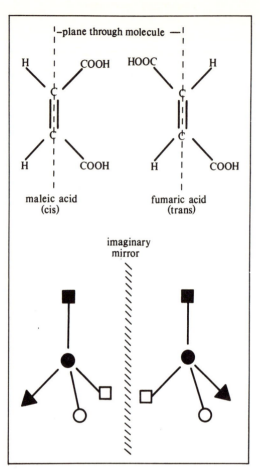

11. *Two forms of isomerism important in biochemistry.* In geometric isomerism (*top*) an imaginary plane is drawn through the centre of the molecule, and the different isomers have their functional groups on different sides of this plane. This form is usually called cis-trans isomerism. If a carbon atom has four different atoms or groups attached to it (*bottom*), it will exhibit optical isomerism. The two isomers are mirror images of each other, and whichever way they are turned or rotated, they cannot be superimposed on each other so that all four groups line up

replaced, the resulting molecule will be the same. If two of the hydrogen atoms are replaced by chlorine atoms, only one type of molecule is possible again, as the corners of the tetrahedral which these hydrogen and chlorine atoms occupy are all adjacent to each other, so that every chlorine atom has one hydrogen and one chlorine atom as its immediate neighbours. Even if one hydrogen atom of methane is replaced by a chlorine atom, and another by, say, a bromine atom, only one molecule is possible.

However, when a carbon atom is bonded to four different atoms or functional groups, a chlorine, a hydrogen, a carboxylic group and another carbon for example, a very different situation arises. For the carbon is now said to be 'asymmetrically' substituted, and there are two possible arrangements in space of these four groups; one arrangement is a mirror image of the other, and they can be seen to be different compounds by the fact that they cannot be superimposed on each other as can molecules with only one or two different attached atoms. This can be demonstrated by oranges and cocktail sticks again. If two models of methane are constructed and placed side by side as though one is a mirror image of the other, each pair of corresponding oranges can be marked by a different colour paint; the two top oranges the same colour, the two 'front' oranges another colour, the two back oranges a third colour. Whichever way these models are now turned, it is impossible to superimpose one on the other so that all the colours line up.

These two models represent two optical isomers of the molecule. They are called enantiomers, and isomers of this type have the same chemical and physical properties—the same boiling point, melting point, solubility and so on. The two forms can only be distinguished from each other by the way they affect the behaviour of polarized light. Normal light is made up of vibrations in all directions, while polarized light consists of vibrations in just one plane, called the plane of polarization. Such light is produced by passing normal light through a special filter which only lets through light vibrations in one plane, cutting out all the rest. If polarized light, or normal light passed first through a polarizing filter, is passed through a solution of one of the enantiomers, its plane of polarization is rotated to the left, while a solution of the other enantiomer will rotate the plane of polarization to the right. These two enantiomers are called respectively the —form and the +form. The actual degree of rotation—how much the plane of polarization is rotated—is dependent on the concentration of the solution of the enantiomer and on the structures of the molecules, so that enantiomers of different compounds will produce different degrees of rotation. Optical isomers will be discussed more fully where they occur in the major classes of biochemicals.

This concludes the background discussion of the principles of organic chemistry and biochemistry. The following chapters will show how these concepts can be applied to the understanding of the small molecules of living organisms, the sugars, the fatty acids and the amino acids, and how these small 'building blocks' are joined together to form huge molecules called polymers of the small molecules. These polymers form the giant molecules which are the real working materials of the biochemist, molecules which the pure organic chemist is only slightly, if at all, concerned with. The definition of a giant molecule varies, and every scientist has his own idea, but in general they are molecules which have a molecular weight in excess of 10,000. In fact as we shall see, a molecule of this size would be very small fry indeed for the biochemist, and some of the polymers have molecular weights of several millions. It is this vast size, and the extremely complex structural organization which this size implies, which make biochemistry such a fascinating subject.

3
The Carbohydrates

In sheer amount the carbohydrates make up the bulk of all the organic substances on the earth. Carbohydrates, the materials we know in everyday life as sugar and starch, are predominantly of plant origin, but they form the principle component of the food materials of most animals, either obtained from plants themselves or indirectly from other animals. Among all the major constituents of our everyday food, carbohydrates supply the majority of the energy required by living organisms, and they have an important part to play in the structures of living systems.

Carbohydrates were first discovered before the importance of the structural aspects of molecules was recognized, and because they were found to contain only the elements carbon, hydrogen and oxygen, with the hydrogen and oxygen in the same ratios as they exist in water, these compounds were thought of as hydrates carbon, and assigned a general empirical formula $C_x(H_2O)_y$, where x and y were different whole numbers. It became clear later on that this definition was unsuitable, as other compounds were discovered which, although they had chemical and physical properties similar to those of the carbohydrates, did not contain the elements hydrogen and oxygen in the ratio 2:1. The compound deoxyribose is such an example, having an empirical formula $C_5H_{10}O_4$. Also some compounds were discovered which did have the elements hydrogen and oxygen in the same ratio as water, but did not show the properties of carbohydrates; lactic acid has the empirical formula $C_3H_6O_3$, but is not a member of the carbohydrates. Later still compounds containing, apart from carbon, oxygen and hydrogen, the elements nitrogen and sulphur, were found to have characteristics which qualified them for inclusion in the class of carbohydrates, and so the original definition became even more inadequate. The present-day position is that carbohydrates and some related compounds are classified according to their general structures, and the original definition has become convenient rather than accurate.

Monosaccharides

The simple carbohydrates, or simple sugars, are called monosaccharides, and these are molecules which cannot be broken down into smaller carbohydrate units under reasonably mild chemical conditions. More complex carbohydrates, which will be described later in this chapter, consist of chains of monosaccharides joined end to end; if there are just a few monosaccharides joined together in a chain, the carbohydrate is called an oligosaccharide, while the polysaccharides consist of large numbers of monosaccharides linked together.

Chemically the monosaccharides are either polyhydroxyl aldehydes, called the aldoses, or polyhydroxyl ketones called the ketoses, that is they are molecules consisting of chains of carbon atoms containing a carbonyl functional group on one of the carbons and hydroxyl groups on, usually, all the rest. The monosaccharides can also be subdivided according to how many carbon atoms they contain; the simplest ones contain just three carbon atoms and are called the trioses, while larger molecules are called tetroses, pentoses and hexoses if they contain four, five or six carbon atoms respectively in their structures. Higher monosaccharides are recognized with seven, eight or even nine carbon atoms, but these are not so important biochemically. In all about seventy monosaccharides are known, some twenty of which are naturally occurring and the rest synthetic, and in fact it is only a few of the naturally occurring sugars which interest us biochemically. They have the general properties of being sweet to the taste, very soluble in water, but only very slightly, if at all, soluble in organic solvents such as alcohol or ether.

Isomeric forms

The simplest monosaccharides, with just three carbon atoms, are the aldose glyceraldehyde and the ketose dihydroxyacetone, shown in diagram 12. From these trioses, the longer chain aldoses and ketoses can be built up (at least on paper—the biochemist and the organic chemist have slightly more complex ways of making carbohydrates) by adding the group HCOH between the carbon atom of the carbonyl group and other carbon atoms in the chain, so that each time the group is added, the carbon chain becomes one atom longer.

CHO	CHO	CH$_2$OH
H——C——OH	HO——C——H	C══O
CH$_2$OH	CH$_2$OH	CH$_2$OH
D–glyceraldehyde	L–glyceraldehyde	dihydroxyacetone

12. *The simplest carbohydrate compounds*, glyceraldehyde containing an aldehyde functional group and so forming the basis of the aldoses, and dihydroxyacetone with a ketone group forming the basis of the ketoses.

If aldoses are written in this way, with the aldehyde group at the top, all the members of the D series of optical isomers have the hydroxyl group on the lowest 'asymmetrical' carbon atom written on the right. If the hydroxyl group is on the left, the molecule is a member of the L series. Dihydroxyacetone shows no optical activity, as it has no 'asymmetrical' carbon atom, and only ketoses with four or more carbon atoms can be present in optical isomers. Again the position of the hydroxyl group on the 'asymmetrical' carbon atom furthest from the carbonyl group determines membership of the D or the L series

It will be noticed that carbon atom number two of glyceraldehyde is 'asymmetric'—it has four different groups attached to it and so its mirror image cannot be superimposed on it. Glyceraldehyde therefore exists in two possible optical isomers. By convention monosaccharides are written with the aldehyde group at the top, on carbon number one, and the —CH$_2$OH terminal grouping at the bottom, so that in glyceraldehyde the two optical isomers are distinguished by whether the hydroxyl group on the middle carbon atom is written on the left or the right side of the molecule. If it is written on the right, the molecule is said to have dextrorotatory configuration (D) and if it is on the left of the molecule written in this way it is called laevorotatory (L).

The configuration of D-glyceraldehyde forms the basis of the classification of all biochemicals which exist in optical isomer form (not only the monosaccharides but also other compounds which will be met in later chapters). If, say, a pentose is written according to the convention with the aldehyde group at the top, it is the configuration of the bottommost (or highest numbered) 'asymmetric' carbon atom which determines to which series, D or L, it belongs—if the hydroxyl group is on the right it is a D-pentose and if it is on the left it is an L-pentose. It must be stressed at this point that the fact that a particular molecule is assigned to either the D or the L series does not give any information whatsoever about which way it will rotate the plane of polarized light. Quite by chance it was found that D-glyceraldehyde rotates the plane of polarized light to the right, and hence the L isomer rotates it to the left, but many compounds assigned to the D series rotate polarized light to the left, and many members of the L series rotate it to the right. To indicate both the absolute configuration of a molecule in relation to the D-isomer of glyceraldehyde, and the direction in which it rotates the plane of polarized light, it is necessary to use both D and L symbols together with $+$ and $-$ symbols. Thus D-glyceraldehyde is fully written $D(+)$-glyceraldehyde and the L form $L(-)$-glyceraldehyde, but other molecules may be written as, say, $D(-)$-pentose, or $L(+)$-hexose.

It is usual to find that living systems can synthesize and use only the members of one series of particular compounds. Among the monosaccharides, only those with the D configuration are found in living organisms, not those with configurations based on L-glyceraldehyde. Even of the D monosaccharides only a few are at all common, but from these, examples which rotate the plane of polarized light to either the left or the right are found.

Each time the carbon chain length of monosaccharides is elongated by the addition of an HCOH group between the carbonyl group and the rest of the carbon chain, a new 'asymmetric' carbon atom is introduced, and therefore the larger monosaccharides exist in many more isomeric forms. In the tetroses derived from glyceraldehyde, there are two such carbon atoms, and thus there are four possible isomers, two D forms and two L forms, each pair of L and D forms having different configurations of the hydroxyl groups and hydrogen atoms on the number two carbon atom. In the pentoses built up from glyceraldehyde there are eight possible isomers, and in the hexoses, 16 forms are known to exist. Generally the number of isomers is indicated by the expression 2^n, where n is the number of 'asymmetric' carbon atoms, so that in the hexoses, with four centres of asymmetry, there are 2^4, or 16 isomers.

When the structures of the monosaccharides are written according to our convention, different isomers can be distinguished according to the positions of the hydroxyl groups on the carbon atoms other than the lowest one, which of course is still used to assign the molecule to the D or the L series. Thus there are, in the hexoses,

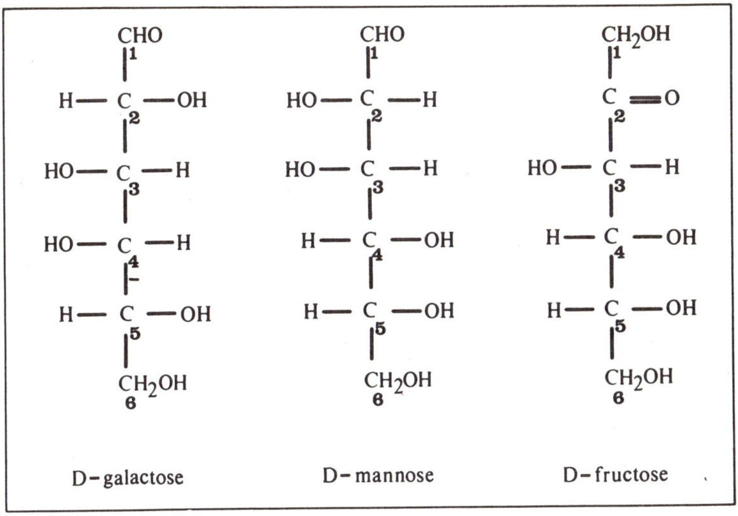

	CHO $\overset{|}{\underset{2}{C}}$			CHO $\overset{|}{\underset{2}{C}}$			CH_2OH $\overset{|}{\underset{2}{C}}$	

The structures shown:

D-galactose

CHO
H — C₂ — OH
HO — C₃ — H
HO — C₄ — H
H — C₅ — OH
CH₂OH (6)

D-mannose

CHO
HO — C₂ — H
HO — C₃ — H
H — C₄ — OH
H — C₅ — OH
CH₂OH (6)

D-fructose

CH₂OH
C₂ = O
HO — C₃ — H
H — C₄ — OH
H — C₅ — OH
CH₂OH (6)

13. *Three common aldohexoses.* The L isomer of each of these compounds would be written with the hydroxyl group on carbon number 5 on the left, in contrast to its position on the right in these D structures

eight pairs of compounds with different configurations on the other carbon atoms, and two of these compounds are shown in diagram 13. The D and L members of each pair have similar chemical and physical properties, and only differ in their behaviour with respect to polarized light, but the differences between the structures of different L and D pairs of isomers are sufficient to endow them with different melting points, boiling points, solubilities and so on. Thus mannose and galactose are different monosaccharides chemically and physically, and because they are not related to each other as enantiomers, they are called diastereomers. D-mannose is an enantiomer of L-mannose, but is a diastereomer of D-galactose and of the other hexoses derived from glyceraldehyde. This type of relationship is quite general throughout biochemistry; if two optical isomers are not enantiomers, then they are called diastereomers.

Dihydroxyacetone, as can be seen from its structure in diagram 12 on page 43, has no 'asymmetric' carbon atom, and so it is not until a HCOH group has been added into the molecule, making a tetrose, that optical isomers are possible, and not until the carbon chain contains five carbon atoms that diastereomers exist. Thus in the ketopentoses there are four possible optical isomers, two D forms and two L forms, and in the ketohexoses, there are eight different isomers. The D and L forms of the ketoses are again distinguished on paper by the configuration of the hydroxyl group on the 'asymmetric' carbon atom furthest from the carbonyl group, which in these molecules is on carbon atom number two. Of the ketoses, we need only be concerned with one of any biochemical significance, a ketohexose called fructose.

Cyclic structures

The straight chain structures which have been used for the monosaccharides so far, although most convenient for showing isomeric forms and the general structural organization of the molecules, do not give a true picture, and these structures do not account for all the observed properties of these substances. For example, although monosaccharides have been shown as aldehydes, they do not in fact show the characteristic chemical properties of simple aldehydes. Furthermore two crystalline forms of D(+)-glucose are known, and there is no way, using merely the straight chain formulae, that these different crystalline forms can be distinguished. That these two forms are different is shown by their effect on polarized light. One form, when dissolved in water, produces a specific rotation of the plane of polarization of 113° to the right when freshly prepared, but on standing this specific rotation will change to 52.5° to the right. The other form, when freshly prepared in solution in water, has a specific rotation of only 19° to the right, but again, after this solution has been standing for some time, its rotation changes to 52.5° to the right. This phenomenon of shift of degree of rotation is called mutorotation, and indicates the existence of two forms of D(+)-glucose. When a solution of one form is made in water, some of the molecules will change to the other form until there is a final solution which is a mixture of the two forms, so that the degree of rotation of the mixture is influenced by both forms.

In nature, pentoses and hexoses exist largely as cyclic structures, and only a very small proportion exists freely as the open chain molecules of the structures we have been using. The formation of the ring structure is quite easy to understand if one remembers that the four covalent bonds of the carbon atom have tetrahedral orientation, so that the carbon chain is not a straight structure, but is considerably bent and the two ends of the chain, far from being isolated from each other, are in fact quite close together in space. In glucose, for example, the hydroxyl group of carbon atom number five lies very close to the carbonyl group on the first carbon atom. This closeness of the two groups facilitates the reaction between the carbonyl and the hydroxyl groups leading to the formation of internal hemiacetal, with resultant closing of the ring. This type of reaction is common to all monosaccharides, hence the predominance of the ring structure.

Depending on which hydroxyl group on which carbon atom participates in the internal hemiacetal, rings can be formed with four carbon atoms and one oxygen atom, a five-membered ring, or with five carbon atoms and an oxygen atom, a six-membered ring. These structures are called furanose and pyranose rings respectively. In the pentose monosaccharides, the furanose ring is usually formed, while among the hexoses, although the furanose ring form may appear in some monosaccharide units when they are joined together in polysaccharides, most of the sugars of interest to biochemists exist in the six-membered pyranose form. The formation of seven-membered rings, which would involve a reaction between the carbonyl group and the hydroxyl group on the sixth carbon atom of a hexose, does not readily occur, as the considerable strains set up in some of the bonds of such a ring would make it very unstable.

A way of writing ring structures of monosaccharides was proposed in 1927 by Haworth. In his formulae, the plane of the ring is considered to be perpendicular to the plane of the paper, such that, in glucose, carbon atoms numbers two and three are in front of the paper, and carbon atom number five and the oxygen atom in the ring are behind the paper. The hydrogens and the hydroxyl groups attached to the carbon atoms now lie either above or below the plane of the ring. Because the hydroxyl group which determines whether a molecule is a member of either the D or the L series is involved in the hemiacetal, and is thus a member of the ring, the configurations of ring structures are recognized by the position of the CH_2OH group, the sixth carbon atom which is not involved in the ring. It is quite valid to use this group to recognize D or L monosaccharides, as it is joined to the same carbon atom which determined the configuration in the straight chain formulae; when this group is written above the plane of the ring the compound is a D-monosaccharide. Different diastereomers are recognized by which hydroxyl groups are above or below the ring.

14. Because of the tetrahedral shape of carbon atoms, a long chain is not a straight structure, but is bent. In monosaccharides, the two ends of the molecule come close together, and the hydroxyl group on carbon number 5 can react with the carbonyl group on carbon number 1 to form a hemiacetal, resulting in the formation of a ring structure (*top*). Such rings can be represented by Haworth structures (*bottom*).

The ring is in a plane perpendicular to the plane of the paper, and the thick lines indicate that this half of the ring is in front of the paper. Carbon atoms in the ring are omitted, but are assumed to occupy the corners.

All D isomers are written with the CH₂OH group on carbon number 5 above the ring, and the different compounds are recognized by the positions of the hydroxyl groups on the other atoms, either above or below the ring. The α and β forms are recognized by the position of the hydroxyl group on carbon number 1

Haworth ring structures of glucose are shown in diagram 14. For clarity, the carbon atoms in the ring are omitted, but they are assumed to occupy positions at the corners of the hexagon ring. The thickened lines indicate that these parts of the ring are in front of the paper, and the groups written above and below the carbon atoms on the paper are in fact above and below the plane of the ring in the actual molecule.

The formation of the ring structures creates a new 'asymmetric' centre in the molecule. One of the structures in diagram 14 shows the hydroxyl group on carbon atom number one at the right to be above the ring, while in the other structure it is below the ring, and these two structures represent the two crystalline forms of D(+)-glucose mentioned earlier. The molecule with the hydroxyl group on this carbon below the plane of the ring is called the α-form, and the molecule with this hydroxyl group above the ring is the β-form, and so the full description of these molecules is α-D(+)-glucose and β-D(+)-glucose.

Some common sugars

The names and structures of a few monosaccharides have already been mentioned, but it will be useful to describe them, and a few others which are common and important in living material, in a little more detail. Of the pentohexoses, ribose and deoxyribose are very important as constituents of the biochemical compounds called the nucleotides and the nucleic acids which are the subject of a later chapter. Both of these sugars occur in the furanose ring form, and deoxyribose differs from ribose only in the lack of a hydroxyl group attached to carbon atom number two in the ring. Other pentoses which are found in nature are called arabinose and xylose, which, however, are rarely found as free sugars, but as components of polysaccharides in various vegetable gums.

Four hexoses are important in biochemistry, one ketohexose and three aldohexoses. Glucose is the most important and widely distributed of all the monosaccharides, and is sometimes called 'grape sugar' or 'dextrose'. It is found free as discrete molecules in sweet fruits and is the basic unit on which most of the polysaccharides, involved as structural components of living organisms and as stores of energy, are built up. It thus forms the bulk of the food material of living things. D(+)-mannose differs from glucose structurally by having different configurations of the hydrogen and the hydroxyl group at carbon number two in the pyranose ring. Although it can occur as a free molecule, it is more commonly found bound to other monosaccharides in long polysaccharide chains. D(+) galactose, the third aldohexose, does not usually occur free, but again in combination with glucose molecules in polysaccharides. It differs from glucose in its configuration at carbon atom four, so that, unlike glucose and mannose, three of its hydroxyl groups lie above the plane of the pyranose ring. Galactose is interesting in that the L form does occur in nature in some vegetable gums, but this molecule cannot be used as food material by other living organisms.

The most common ketohexose in nature is called fructose, and although it has the ability to rotate the plane of polarized light to the left, it is a member of the D series of ketohexoses, having the same configuration at its lowest carbon atom as D-glucose, and is thus written D(−)-fructose. In its free form, which occurs in honey and some fruits, it normally exists in the six-membered pyranose ring structure formed by an internal hemiacetal between the ketone group on carbon atom number two and the hydroxyl group on the sixth carbon atom. But when it is combined with glucose in the oligosaccharides or polysaccharides, it is more often present in the furanose form, the hydroxyl group on carbon atom number five being involved in the hemiacetal formation and ring closure.

Chemical reactions of the monosaccharides

The chemical reactions in which monosaccharides participate are determined mainly by the hydroxyl groups, and few reactions of the carbonyl group are seen. This is because the proportion of molecules with free carbonyl groups is very small, the aldehyde and the ketone groups usually being involved in the hemiacetal. However, reduction of the carbonyl group gives rise to the corresponding pure poly-alcohol. For example, reduction of the aldehyde group in mannose results in the formation of the alcohol called mannitol.

The few free carbonyl groups which do exist in a sample of a monosaccharide are responsible for the observed properties of the monosaccharides as reducing agents, these properties being due to the carbonyl group adjacent to a hydroxyl group in the molecule. Reducing properties of sugars can be used to test for their presence in a sample in the laboratory; for example, if a solution of a silver salt is mixed with a sugar solution the silver will be reduced (reduction is the addition of electrons to an ion, so that the silver ion Ag^+ is reduced to the silver atom Ag) and a thin deposit of silver will be produced on the sides of the glass test-tube. This silver deposit gives the test-tube sides a mirror-like appearance—hence the name of this test, the 'silver mirror test'. Solutions containing copper ions (Cu^{++}) will also be reduced by monosaccharide solutions, the copper ions being reduced to cuprous ions (Cu^{+}), which are insoluble in basic solutions, so that a brick red solid is produced in the test-tube. This reaction forms the basis of the test known as Fehling's reaction.

Reactions of the hydroxyl groups give rise to a number of important derivatives of the simple monosaccharides. Oxidation of the terminal $-CH_2OH$ group to a carboxyl group, without oxidation of the aldehyde group, yields compounds called the uronic acids. Thus glucuronic acid, the oxidation product of glucose, is the basic compound of this series, and is important because it combines very easily with other monosaccharides to form digosaccharides and polysaccharides. Many substances are excreted in the urine of mammals only after they have been coupled to glucuronic acid.

Hydroxyl groups in monosaccharides can undergo chemical reactions similar to those in the simple organic alcohols, and among the most important are those reactions leading to the formation of esters, in particular esters with phosphoric acid—the sugar phosphates. For example, esterification of the hydroxyl groups on carbon atoms numbers one or six in glucose yields glucose-1-phosphate or glucose-6-phosphate respectively, and these compounds play a vital role in the metabolism of the carbohydrates in living systems. In some way, esterification with phosphoric acid seems to 'activate' the sugars to take part in the chemical reaction of metabolism. Although these two phosphates of glucose are derivatives of the very same sugar, they each have different properties and different roles and functions in biochemistry.

The pentoses, ribose and deoxyribose, are always present in the nucleic acids and the nucleotides as phosphate esters. Ribose can, and indeed does, form phosphate esters through its hydroxyl groups on carbon atoms numbers two, three and five, while deoxyribose lacks a hydroxyl group on carbon two, and can thus form only the compounds deoxyribose-3-phosphate and deoxyribose-5-phosphate.

Replacement of one of the hydroxyl groups in a monosaccharide by an amine group leads to another series of sugar derivatives called the amino sugars. D-glucosamine is related structurally to D-glucose, has its amine group attached to carbon atom number two in the pyranose ring, and is the basic member of the series. It is found combined with other monosaccharides in some of the polysac-

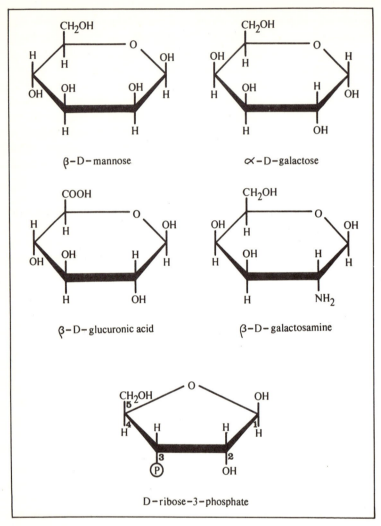

15. *Haworth structures of some common monosaccharides*. All these compounds are members of the D series because their CH₂OH groups are above the plane of the rings. Ribose is a pentose monosaccharide, with only five carbon atoms, so that it normally forms a furanose, six-membered ring

charides, as is another important derivative galactosamine, in which the hydroxyl group on carbon number two in the galactose molecule has been replaced by an amine group. Examples of some of the important derivatives of the monosaccharides can be seen in diagram 15.

Oligosaccharides

A typical chemical reaction of hemiacetals is that they can react with the hydroxyl groups of alcohols to form the full acetals, or simply acetals. The hemiacetals in the cyclic structures of the monosaccharides are no exception, and at once this reaction shows a way in which monosaccharides can be linked together through the hemiacetal of one sugar and a hydroxyl group of another. This is just what happens.

The acetals of monosaccharides are called generally the glycosides (acetals of glucose are often given their own special name, the glucosides, but for simplicity we shall refer to all carbohydrate acetals as glycosides), and so the sugars are said to be linked together by glycoside linkages. Depending on the number of monosaccharide units which are joined together by these glycoside linkages, the complex higher carbohydrates are called either oligosaccharides or polysaccharides. Oligosaccharides (from the Greek *oligos*, a few) are composed of between two and eight monosaccharide units (some scientists consider an oligosaccharide to be composed of up to six units, others up to ten or twelve units—in fact there is no hard and fast definition), and are further sub-divided into the disaccharides, with just two sugars joined together, the trisaccharides with three and so on. Carbohydrates with more than eight (or six or ten!) monosaccharide units are called the polysaccharides, and in fact the total number of units in some carbohydrates can be many hundreds or even thousands. In all cases, whether two or two thousand monosaccharides are joined together, they are all present in their ring form.

Glycoside linkages then are formed between carbon atom number one, the carbonyl carbon atom (or carbon atom number two in the ketoses) and a hydroxyl group on carbon atoms numbers two, three, four or six of another monosaccharide molecule. Because there are two possible configurations of the hydroxyl group on carbon one, giving the molecule either α or β configuration, there are two forms of glycoside, again either α or β depending on the configuration at carbon atom one. But unlike the α and β forms of glucose and other sugars, α and β glycosides cannot interconvert—they are either one form or the other.

By far the most important oligosaccharides biochemically are the disaccharides, which like the monosaccharides are crystalline, soluble in water and have a sweet taste. They can be divided into two types. If one of the monosaccharide molecules in the disaccharide has a free hydroxyl group joined to the carbonyl carbon atom, as it would have according to the above description of the glycoside linkage, then the disaccharide will still show reducing properties, and will give positive reactions in the 'silver mirror test' and in Fehling's reaction. The simplest disaccharide of this type is a compound found in malt and hence called maltose, or malt sugar, and so all of the reducing disaccharides are said to be of the maltose type. Another consequence of the free hydroxyl group at carbon number one is that these disaccharides will show the phenomenon of mutorotation, just as in the two crystalline forms of D-glucose.

The maltose molecule is composed of two glucose units joined together by an α-1-4-glycoside linkage, which means that the hydroxyl of carbon number one in one glucose is in the α configuration, and is joined to the other glucose by linking to the hydroxyl group on carbon atom number four. Two glucose molecules linked together by α-1-6-glycoside linkages form a disaccharide called isomaltose, a substance which in its free state is not very important, but which is involved in the structures of some of the polysaccharides.

Two other reducing disaccharides are of interest to us here. Cellobiose consists of two glucose molecules joined by a β-1-4-glycoside linkage, the hemiacetal hydroxyl group involved in this linkage being in the β configuration and linked to the other glucose unit by the hydroxyl group on the carbon number four as in maltose. Lactose is also formed from two monosaccharide units joined by a β-1-4-linkage, but in this case, a glucose molecule is joined to the galactose molecule, and it is the hemiacetal hydroxyl group of the galactose which is involved in the linkage to the hydroxyl group on carbon atom four of the glucose. Lactose is found in large quantities in the milk of mammals—human milk contains about six per cent lactose and cow milk about five per cent—and it can be made commercially as a by-product in the manufacture of cheese.

There is no reason why the hydroxyl group on the carbonyl carbon of one monosaccharide should not form a glycoside linkage with the hydroxyl group on the carbonyl carbon of another monosaccharide instead of one of the other hydroxyl groups, and there is

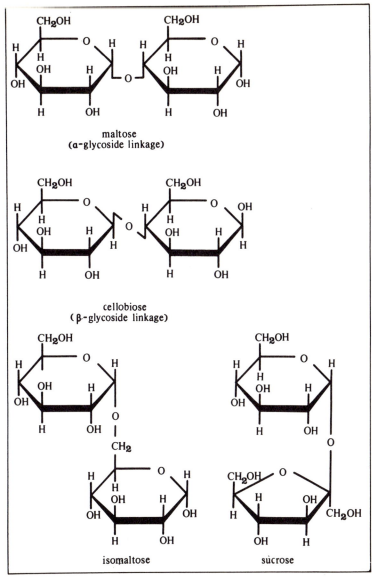

maltose
(α-glycoside linkage)

cellobiose
(β-glycoside linkage)

isomaltose

sucrose

16. *Types of glycoside linkages in disaccharides.* In both *maltose* and *cellobiose*, one glucose unit is joined through its carbon number 1 to carbon number 4 of the other, but in maltose the linkage is at the α position, while in cellobiose it is through the β position. In *isomaltose* the linkage is between carbon number 1 of one glucose and carbon number 6 of the other, and this is also an α linkage. *Sucrose* is composed of one glucose unit and one fructose unit in the furanose ring form. In *maltose, cellobiose* and *isomaltose*, there is a hydroxyl group free attached to the carbonyl carbon atom (carbon number 1) so that these molecules show reducing properties. *Sucrose* has no such hydroxyl group, and hence is a non-reducing compound

a very important disaccharide which has just this type of linkage. It is called sucrose, and it is found widely throughout the plant kingdom, although only the sucrose in sugar beet and sugar cane is used commercially as the 'sugar' which we put into our tea or coffee. Sucrose is probably the only common food which is used in its crystalline form.

Because both of the hemiacetal hydroxyl groups of the two monosaccharide units are involved in the glycoside linkage in sucrose, this sugar has no reducing properties and does not show mutorotation. It is formed from a glucose molecule and a fructose molecule, and it is interesting that the fructose unit is present in the five membered furanose ring, and not in the pyranose structure which it shows when it is a free molecule. Sucrose is strongly dextrorotatory—it rotates polarized light to the right—but when the constituents of this disaccharide are split apart, the mixture becomes laevorotatory, mainly under the influence of the fructose. This splitting is therefore called inversion, and the resulting mixture, invert sugar, is commonly found in honey together with sucrose itself.

Polysaccharides

Because oligosaccharides of the maltose type have a free hydroxyl group on the carbonyl carbon atom of one of their monosaccharide units, more monosaccharide units can be linked together to form chains with many more units, and such chains are the polysaccharides, the carbohydrates which are of real interest to the biochemist. Polysaccharides are very widely distributed in nature, where they have functions in the structural make-up of living organisms or as storage compounds for food materials. The role of a polysaccharide as either a structural or a storage compound is determined largely by which monosaccharide units are involved in the molecular organization, and by the specific ways these units are linked together by glycoside linkages.

Polysaccharides can be composed of many thousands of monosaccharide units, and hence they have extremely large molecular weights. Determining these molecular weights is rather difficult, and in fact a definite molecular weight often cannot be assigned to a polysaccharide, because the length of the chain is not a definite exact number of units. For example, in a chain consisting of several thousand units, the addition of 10, or 50 or even 100 units does not seem to make much difference to the chemical and physical properties of the whole molecule, so that a given polysaccharide can exist in chains of varying lengths. It is useful, however, to obtain an estimate of the molecular weights. One way of doing this is to find out how many units are involved in the molecule, and then to add together the molecular weights of all the constituent units. This is a laborious method, and so other physical methods are normally employed for finding molecular weights of such large compounds.

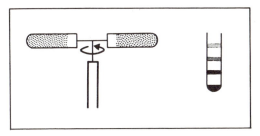

17. The use of the centrifuge for determining molecular weight. After the samples have been spun for some time, all the particles separate out into discrete bands, the constituents of each band being a different weight, the heaviest at the bottom. The position of each band is a measure of the rate at which its constituents have fallen to the bottom during the spinning, and from this value the molecular weight can be estimated

In the case of the polysaccharides, the most useful method is to observe the behaviour of the molecules when spun in a centrifuge. A centrifuge is a piece of equipment which enables the scientist to subject samples to very high centrifugal forces. If a bucket of water is swung round and round in a circle, the water will not spill out, even while the bucket is upside down at the top of the circle, because the water is being forced towards the outside of the circle, or the bottom of the bucket, by centrifugal forces; the same principle is used in the centrifuge. In scientific instruments, however, the speed of rotation is enormous, so that the solutions being spun are forced even more strongly towards the outside of the circle. Any particles in the solution will also be forced to the outside of the circle, or the bottom of their container, and the rate at which these particles collect on the outside is a function of their weight. Thus with a mixture of particles of different weights, the heaviest will be forced right to the outside of the circle, and the lighter particles will form layers progressively towards the inside. The scientist can therefore observe how far the particles of the polysaccharides have been forced to the outside, and this provides him with a value for their weights, and thus for their molecular weights.

Because of the way in which they are made up, the number of possible different polysaccharides is astronomical. The molecules can be made up of all or any of the monosaccharide units in any order, they can be joined together by any of the points of attachment on the individual units, that is to the hydroxyl group on carbon atoms numbers two, three, four or six, and by linking a free carbonyl hydroxyl group of a sugar already forming part of a chain to another glucose already part of another chain, there is the possibility of branched chains, and these branches can themselves have branches. So the number of possible permutations in the polysaccharides is somewhat greater than the number of possible permutations on an average football pools coupon. In reality, however, the situation is a great deal simpler, as only a small number of all the possibilities actually occur, and even among the naturally occurring polysaccharides, small variations seem to make little difference to the properties of the compound. In biochemistry, we need only be concerned with the structures and functions of three or four polysaccharides.

One aspect of the structure of the polysaccharides which the biochemist does need to know in some detail is what monosaccharide units are present. He can find this out by breaking up the chains and then analysing the mixture of monosaccharides obtained. Because the formation of the glycoside linkage resulted in the loss of a water molecule between the two monosaccharides, the linkage can be broken by giving back this water, a process called hydrolysis. (We will meet this reaction later as a method of breaking any bond formed by the loss of water, such as ester linkages.) Sometimes the addition of water is very difficult, and it has to be 'forced' in; so the usual method is to treat the carbohydrates with very strong acids, such as hydrochloric acid, to break apart the glycoside linkages. If the polysaccharide is composed only of one type of sugar unit, then the job of analysing it is almost complete. Such polysaccharides are called homoglycanes. Another group, called the heteroglycanes, contains two or more different monosaccharide units, and with these the task of analysis is a little more difficult, for not only must the identity of the units be determined, but also in what order and by what linkages they are joined together. So the biochemist employs special chemicals, called enzymes, which are able to hydrolyse and so break up the polysaccharide chain at specific points, for example at a glycoside link between a glucose and a galactose unit. Once the chain is broken down into these smaller fragments, the job of analysing these is of course much simpler. If the polysaccharide is very complex, with many different monosaccharide units joined in different ways, a number of enzymes can be used, each of which will break up the chain at different points and will break different glycoside linkages. By repeating this type of process, different samples of the same polysaccharide can be broken down in different ways, to give different smaller fragments, and when each of these has been analysed, the structure of the complete chain can be deduced from all the results. But solving this puzzle can itself be very time-consuming, and in recent years computers have been used in the final reconstruction of the chain. Although this sounds a very complex process, it is simplified by the fact that the common polysaccharides contain just the monosaccharides glucose, the most common constituent, sometimes together with fructose, galactose, and other hexoses.

Cellulose

By far the most important structural polysaccharide is cellulose, a major component of plants, where it is found chiefly in the rigid walls surrounding the plant cells. In some plant materials it forms the majority of the structure; cotton and flax for example contain between 90% and 99% pure cellulose, and wood is composed of some 45% cellulose. In fact it has been estimated that there is more cellulose in the world than any other organic chemical.

Complete hydrolysis of cellulose yields only D-glucose, showing that cellulose is a homoglycane. But partial hydrolysis yields the disaccharide cellobiose, indicating that cellulose can be thought of as a repeating sequence of units of this disaccharide. All of the monosaccharide units in the cellulose chain are therefore linked together by β-1-4-glycoside linkages, as are the sugars in cellobiose. The molecular weight of cellulose seems to vary a great deal; some scientists report values of 50,000, while others claim that the weight is as high as two million; in any case, the molecule is a very large one, with many thousands of D-glucose units joined together in long straight chains.

The value of the cellulose molecule as a structural component of living tissues is a consequence of its long straight chains. These molecules can be packed together in bundles, each molecule lying in the same direction, to form crystal-like threads which have a strength somewhat greater than a thread of high quality steel of the same thickness. Each bundle in a thread is called a micelle. When the uniform orientation of cellulose is lost and the micelles become shuffled to lie in random order in any direction, the fibres lose their great strength. Commercially this loss of orientation is put to good use in the manufacture of the synthetic substance we call cellophane.

Starch and glycogen

The major storage polysaccharide of the plant kingdom is starch, which is used as a reserve of food materials which the plant can draw on when needed. It is the major component of plant tubers such as potatoes and is present in large quantities in fruits and in seeds, and such structures can be composed of up to 70% starch. Like cellulose, starch is composed solely of glucose units, so that when the plant needs to draw on its food reserves, it merely breaks down some starch and uses the resulting glucose.

But the similarity with cellulose ends here. For starch is present in granules, and there is none of the organized crystalline structure which is such an important feature of the structural polysaccharide. Furthermore starch is composed of two types of glycoside-linked chains, called amylose and amylopectin, of which the latter usually forms some 70% to 80% of starch by weight.

Partial hydrolysis of the amylose component of starch yields mainly maltose disaccharides, so the glucose units in amylose must be linked by α-1-4-glycoside linkages, the same type of linkage found in maltose. Other analyses have shown that the chains of this component are straight and unbranched, and typically contain about three hundred glucose units, although scientists disagree on this figure and there may be slight differences between different samples of starch analysed. The characteristic chemical test for the presence of starch is to mix iodine solution with a sample, when a deep blue colour is

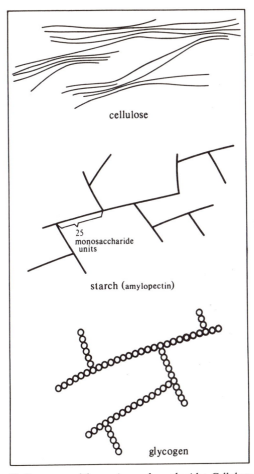

cellulose

25
monosaccharide
units

starch (amylopectin)

glycogen

18. *Structures of the common polysaccharides.* Cellulose is composed of fibres lying close together and forming tight bundles which have great strength. In both amylopectin and glycogen, the chains are branched, every 25 to 30 units in starch and every 10 to 14 units in glycogen, where each circle in the diagram represents one glucose unit

produced, and this colour is produced by the amylose component. Thus if a potato is immersed in iodine solution it will be stained almost completely deep blue, because it contains a great deal of starch. The other component of starch, amylopectin, if mixed on its own with iodine solution will just give a red colour.

Amylopectin is quite different in structure from amylose. Partial hydrolysis yields a mixture of disaccharides, maltose and isomaltose, so that this compound is composed of glucose units linked by α-1-4-linkages and some linked with α-1-6-linkages. Again unlike amylose, amylopectin is not a straight molecule, but is highly branched, and its structure can be pictured as a chain of maltose units joined together, with branches every thirty units or so, and these branches are formed by glycoside linkages to a hydroxyl group on carbon atom number six on one of the glucose units in the chain. Thus the isomaltose disaccharides isolated from partially hydrolysed amylopectin represent the branch points in the molecule. The complete molecules of amylopectin are much larger than those of the amylose component, having molecular weights of up to 500,000 or even more.

A polysaccharide called glycogen is the main storage carbohydrate of the animal world. It is found chiefly in muscle and liver tissues, the liver containing the largest amount of glycogen per unit weight of tissue. Once again it is composed solely of D-glucose monosaccharides, so that the animal, when it needs food materials for its bodily processes, merely mobilizes some of its glycogen store and breaks it down to glucose units.

The molecular weight of glycogen seems to vary between three hundred thousand and ten million, and it has a branched structure similar to that of amylopectin. But the length of the sugar chains between the branches is only about ten to twenty units, so that it is even more highly branched than the starch component. Glycogen is rather interesting among the polysaccharides in that it is soluble in water, while most of the others are completely insoluble.

Other polysaccharides

Cellulose, starch and glycogen are certainly the most important polysaccharides encountered in nature, and are of most interest to biochemists. But one or two others do play important roles, and are worthy of brief mention. A polysaccharide composed mainly of fructose units present in the five-membered ring structures is found as a storage compound in the bulbs of many plants, and is especially common in some plants such as asparagus and Jerusalem artichokes. It is called insulin and is probably responsible for the characteristic taste of these vegetables. Pectins, compounds present in the fruits of many plants, and which are used in the kitchen for making gels, for instance in jam or jelly making, are polysaccharides composed of galactose, arabinose and galacturonic acid joined with α-1-4-linkages.

Although polysaccharides are used extensively as structural materials in the plant kingdom, their use in animals for structural purposes is rather limited. Those that are important usually contain some amino sugars in their structures, and may also contain the element sulphur. Chitin, the black, hard, shiny material which forms the coats of insects such as beetles and the exoskeletons of crabs and related animals, is a polysaccharide composed of repeating units of disaccharides of uronic acids linked to a complex monosaccharide derivative, N-acetyl-D-glucosamine. This molecule is simply gluco-samine in which the amino group has undergone a reaction with acetic acid to form the acetylated derivative. Another polysaccharide which contains this molecule is hyaluronic acid, which is found in the tissues of the eye and in the umbilical cord as a structural compound. With a molecular weight of up to five million, it is composed of repeating units of disaccharides of N-acetyl-D-glucosamine and D-glucuronic acid. An amino sugar in which the $-CH_2OH$ group on carbon atom number six has been esterified with sulphuric acid to form the sulphate ester, is involved in another animal structural polysaccharide, chondroitin sulphate. This is found in cartilage, bone and skin.

Although the possible number of polysaccharides is very large, few are found in nature and so they offer little scope for physical and chemical adaptation. While this lack of adaptation is adequate for structural purposes in the plant kingdom, in the animal world, much more structural flexibility is required. For instance the insects which have hard coats of polysaccharide material are severely limited in the size to which they can grow. So most of the structures of animal tissues are formed from other biochemicals, the lipids and the pro-teins, which enable them to achieve a more flexible and adaptable existence, and the following chapters will describe the structures and functions of these compounds.

4

The Lipids

When we come across the word 'fat' in everyday life, a number of ideas spring to mind. Immediately we think of the fats we use—butter, margarine and lard—and depending on our word usage, we sometimes refer to cooking oils as fats as well. Then maybe we think of some of our friends who are a little overweight, people who are sometimes, rather unkindly, called 'fat', and if we extend this idea we remember that whales and seals have a layer of blubber beneath their skin—another form of fat. Thinking of seals and whales leads us on to Eskimos and we recall that Eskimos are popularly thought to eat a great deal of fat to keep them warm; there is the old joke about the Eskimo who preferred to eat the candles on his birthday cake, rather than the cake itself. So we realize that candles are another form of fat.

After this it becomes more difficult to think of examples. So let us think of the properties of fats. We find that we can say little except that these materials are rather soft, and that they are very water repellent and hence, to remove a fatty stain on clothing, for example, we have to use petrol or carbon tetrachloride. The water repellent property of fats reminds us that the leaves of some plants such as holly and other evergreens have a smooth texture and are very water repellent—water just runs off them—and so if we are really pushed to think of examples of 'fats' we might be tempted to put forward the materials on the outsides of these leaves as candidates.

Although this list does not seem very extensive, it in fact includes examples of most of the main sub-divisions of a class of compounds which the biochemist calls the lipids. The fats, the most abundant compounds in this class, are represented by butter, margarine and lard; the oils by the cooking and frying oils and waxes; and a third sub-division of lipids, by the candles and the shiny coats of leaves. As for the properties, there are indeed few which are characteristic of all members of the lipids, except that they are all composed chiefly of the elements carbon, hydrogen and oxygen, and that almost all are insoluble in water but very soluble in organic solvents such as ether, chloroform and carbon tetrachloride. Further than this we cannot go, for the lipids are a very varied group of biochemicals, and even biochemists find it difficult to decide among themselves on a single standardized name for them. They are thus variously called the lipids, the lipides and even the lipins, but these names all refer to the same compounds, and we shall call them lipids.

The range of lipid compounds

The most convenient way of studying the lipids is to divide them into two broad classifications: simple and compound. The simple lipids embrace the fats, the oils and the waxes which we have already met, and these compounds are esters of an alcohol called glycerol and long chain carboxylic acids. The compound lipids are more complex structures involving, as well as carboxylic acids and alcohols, which may be different from glycerol, other groups containing phosphorus, or complicated organic substances such as carbohydrates or proteins, and this class of lipids includes the phospholipids, the glycolipids and the lipoproteins respectively. There is a class of substances called the steroids, which are usually studied together with the lipids, although they have completely different structures, properties and functions, and so are sometimes called just lipid-like compounds. Some examples of these will be discussed at the end of the chapter.

Simple lipids—the triglycerides

The alcohol involved in fats and the oils—glycerol—is a trihydric alcohol, and thus is able to form esters with one, two or three carboxylic acid molecules. All of these compounds are called generally the glycerides, after the alcohol, and so the esters are called monoglycerides, diglycerides and triglycerides. In the fats and the oils it is almost wholly the triglycerides which interest us. The three carboxylic acids which form esters with glycerol in a particular fat can be either the same or all different, or two of one type and one of another. Their position on the glycerol molecule seems to be random. In a sample of a given fat, molecules will be found where different carboxylic acids occupy either the end positions or the middle position, and this seems to make little difference to the molecule as a whole.

The carboxylic acids we shall be concerned with are all of very long carbon chain length, usually monocarboxylic with the hydrocarbon chain attached to a terminal carboxyl group carbon atom. Because of their initial discovery as constituents of fats, they are more usually called the fatty acids. Although the number of fatty acids existing in nature is very large, the number involved in the lipids is quite small; just three fatty acids comprise about 90% of the fatty acids encountered in the triglycerides of fats and oils, and these have carbon chain lengths of either 16 or 18 carbon atoms. It is characteristic of the glyceride fatty acids that they contain an even number of carbon atoms, a consequence of their synthesis in living organisms, where they are built up progressively by adding two-carbon units to a growing carbon chain. The significance of the numbers 16 and 18 is, however, not clear.

We can divide the fatty acids into two main groups which account for most of the compounds found in the lipids—the saturated and the unsaturated acids, depending on the way the carbon atoms are joined together in the hydrocarbon part of the molecule. And this classification corresponds to the differences in properties between the fats and the oils. Fats are composed solely of saturated fatty acids, while the oils contain one or more unsaturated acids in their structures. Thus the glycerides of unsaturated fatty acids have a lower melting point than those of saturated acids, and the more the fatty acids are unsaturated, that is the greater the number of double bonds in the carbon chains of these acids, the lower will be the melting point.

There is an interesting example of the importance of these structures and their related melting points. A sea anemone called *Metridium dianthus* contains a great deal of fatty material in its tissues, and in species living in warm equatorial waters, these materials are largely triglycerides of saturated fatty acids which are liquid at these warm temperatures. But in related species living in cooler waters, the fatty content is made up more of unsaturated fatty acid triglycerides, which have a lower melting point and are thus liquid at these cooler temperatures. If the equatorial species are placed in the cooler waters, their fatty material will solidify and the creatures will stiffen to the point of immobility.

The saturated fatty acids have the general formula $R-COOH$, where R is the hydrocarbon part of the molecule with the formula

$CH_3(CH_2)_n-$. The number of CH_2- units can range as high as 86, making a fatty acid molecule containing 88 carbon atoms, as in mycolic acid. But by far the most abundant fatty acids in the triglycerides are palmitic acid, with 16 carbon atoms and stearic acid with 18 carbon atoms, and there are only small amounts of shorter

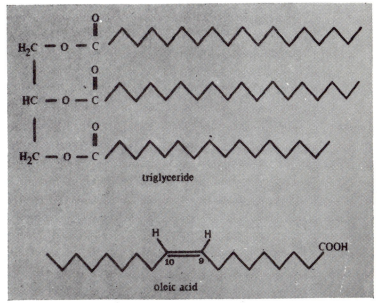

19. *Components of simple lipids.* The triglyceride is a triester of the trihydric alcohol glycerol, and the zig-zag lines represent the hydrocarbon chains of the fatty acids. The upper two each have 18 carbon atoms (a carbon atom is assumed to be at each corner, and is omitted for clarity) and are stearic acids, while the lower one has 16 carbon atoms and is palmitic acid. While both stearic and palmitic acids are fully saturated, oleic acid shown below is unsaturated, and has a double bond between carbon numbers 9 and 10. Such acids are the main constituents of the oils. The hydrogen atoms have been put in on this structure to show that they are in the cis positions

and longer chain acids present in most fats. Most of the common fats are composed of mixtures of glycerides of both of these fatty acids, together with less common ones, but sometimes a single fatty acid predominates. In beef tallow, for example, only stearic acid is involved.

Most of the unsaturated fatty acids found in the fats and the oils have just one double bond in their hydrocarbon chains, and are thus called monounsaturated acids. They can be assigned the general formula $R-CH=CH-(CH_2)_7-COOH$, so that counting the carboxyl group carbon atom as number one, the double bond occurs between carbon numbers nine and ten, but the significance of this position is not clear. Just as with the saturated fatty acids, hydrocarbon chains with 16 and 18 carbon atoms predominate, and the most common are palmitoleic acid with 16 carbons and oleic acid with 18 carbons. And again mixtures of triglycerides with different unsaturated fatty acids are found in the oils, while some are composed mainly of one acid. Olive oil, for example, is found to be mainly triglycerides of oleic acid. Unsaturated fatty acids with more than one double bond in their structures are sometimes encountered, and these are referred to as the polyunsaturated acids. Normally they have their extra double bonds on the opposite side of the original double bond to the carboxyl group, so that the first double bond is still between carbons nine and ten.

Double bonds in the hydrocarbon chain raises the possibility of cis-trans isomerism in the fatty acids. The hydrogen atoms attached to the carbon atoms involved in the double bond can be either on the same side of a plane drawn through the double bond—the cis configuration—or on opposite sides of this plane—the trans form. As with most pairs of stereoisomers, only one form is found in nature, and in this case it is the cis-form of the unsaturated fatty acids which is naturally occurring. •

Most of the fatty acids in living systems, whether saturated or unsaturated, have straight chain hydrocarbon regions of their molecules, but some branched chain acids do exist in triglycerides and recent evidence is indicating that they are probably more common than was once believed. The ones which have been isolated from animal fat samples have odd numbers of carbon atoms, usually between 13 and 17 in the whole molecule.

Converting unsaturated triglycerides into their fully saturated counterparts is a fairly easy process, which has importance in the food industry. The process is called 'hardening' because the liquid, low melting point unsaturated oils can be processed to make solid cooking fats and margarine. Chemically the reaction involved is a hydrogenation, or the addition of hydrogen to the two carbon atoms involved in the double bond, so that this section of the carbon chain, and hence the whole hydrocarbon chain becomes fully saturated. The reaction is really a reduction reaction, but when talking about the addition of hydrogen to carbon double bonds, we usually prefer to use the term hydrogenation. Similarly, oxidation of carbon single bonds to double bonds is called dehydrogenation—the removal of hydrogen.

Chromatography

Biochemists have a number of ways of studying fats and oils to determine their composition. As with the carbohydrates, the first step is usually to break down the molecules into their constituent parts, and in the case of the triglycerides, this means breaking them down to glycerol and a mixture of fatty acids. Breaking the ester bonds between the alcohol and the fatty acids is fairly easily accomplished by hydrolysis with just dilute solutions of acids or bases. If acid hydrolysis is used, the products are free acids, while if basic hydrolysis is employed the products are salts; hydrolysis with dilute sodium hydroxide solution will yield a mixture of the sodium salts of the fatty acids, sodium stearate, sodium oleate and so on. When the fats and the oils have been broken down in this way, there is the problem of separating the different fatty acids and identifying them, so that some idea of the structure of the fat can be obtained.

One of the most useful tools the modern biochemist has at his disposal for separating fatty acid mixtures into their components is a technique called chromatography. Chromatography (the name comes from the Greek, *chroma*, colour) is a technique developed from the observations of a Russian botanist, Tswett, that solutions of pigments from plant material, when allowed to trickle down through a glass column packed with finely powdered alumina (aluminium oxide), separated out into different coloured bands in the column. Each band represented a different single component of the mixture, and so it was an easy matter to cut up the column and to extract the components from each band. Today, the substances which are commonly separated by chromatographic methods are usually not coloured, and the name refers simply to the special separation technique.

Separation of the components of a mixture involves the interaction of each component with two different materials at the same time, one moving in relation to the other. In the glass column packed with alumina, the alumina has the role of the stationary medium, or stationary phase, while the solvent in which the mixture of pigments is dissolved forms the mobile phase. As the mobile phase carries the mixture over the stationary phase, all the components of the mixture come into contact with the fine particles of the alumina, and some of the components actually become attached to and trapped by the powder for a short time. Meanwhile the rest of the mixture has passed on, carried forward by the advancing mobile phase, and so the molecules which have become trapped are left behind, and can only follow the rest of the mixture when they are released by the stationary phase. But no sooner has one region of the stationary phase released them, than the next section traps them, and so they lag even further behind the main body of the mixture. All of these molecules which are trapped are of one type, forming just one of the components of the mixture, and so because they are left behind, they have been separated. As the mixture continues its way along the stationary phase, molecules of another component become trapped by the particles, and so begin to lag behind the rest of the mixture, and they suffer the same fate as the first molecules; when released from the particles, they are trapped soon afterwards. Thus as the mixture travels along the stationary phase, all of its components gradually fall behind, and only follow what is left of the mixture at a slower pace. By the time the mobile phase has reached the end of the stationary phase, it has lost all of the mixture, and the various components are left in the column as distinct bands. If more pure solvent is trickled through, each component will be washed out of the column as a separate solution, or the column can simply be cut up to retrieve the components.

Instead of a glass column, a simpler method of chromatography uses filter paper as the stationary phase. A small spot of the mixture is put on to one end of the filter paper, and the paper is then dipped into a suitable solvent. Just as ink will flow along blotting paper if an edge is placed in a pool of ink, the solvent begins to travel along the filter paper. When it reaches the spot of mixture, it dissolves it and carries it along with it, and the components are separated by just the same method as in the column, the fibres of the paper trapping and holding the components of the mixture. When the advancing front of the solvent has reached the end of the filter paper, the separation will be complete, and to stop all the components from continuing their slow journey and eventually falling off the end of the paper, it is dried quickly and the components can be seen as discrete spots at varying distance from the original spot. Sometimes these spots are invisible, and the paper needs to be developed, perhaps by staining them, or by viewing them under ultraviolet light, depending on the compounds involved.

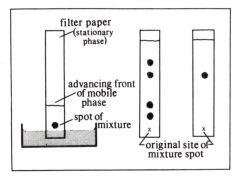

20. *Paper chromatography used to separate a mixture into its components.* The left diagram shows the mobile phase travelling up the filter paper, and the two other diagrams show completed separations. The extreme right diagram shows just one spot, which corresponds with one of the spots from the mixture, so that the spot from the mixture can be identified as the same compound present in the single spot

Once the mixture has been separated, it is a simple matter to identify the components: the distance which each one has travelled from its original position as the spot of mixture is characteristic for a particular compound, provided all the other conditions such as total distance travelled by the solvent, type of solvent used and type of stationary phase employed are kept constant. So by measuring the distance which the component has travelled, and comparing this with the distance which a known compound has travelled under identical conditions, a spot or band can be identified.

Chromatographic techniques using solid stationary phases and liquid mobile phases have been found very useful for separating mixtures of different lipids. But for separating mixtures of fatty acids into their components, a newer technique is of much more value. Instead of a solid stationary phase, a liquid is used, coated on to the inside of a thin glass tube. And instead of a liquid mobile phase, a stream of gas is used, passed through the tube, or column, under constant pressure. This type of chromatography is thus called gas-liquid chromatography. But even with these differences in materials, the mechanism of the separation is essentially the same. The mixture is introduced into the gas stream, and is carried with it along the column. The components interact with the liquid stationary phase, and are trapped and hence slowed down by it. So each component emerges from the end of the column at a different time, and by comparing this time between introduction of the mixture and the emergence of a component with the time taken by a known compound under identical conditions, each component can be identified. The normal practice is to put a detector at the end of the column, which will signal the emergence of each component by an electrical impulse, and these impulses can be recorded automatically by a pen moving over a revolving drum or moving sheet of paper. Modern gas-liquid chromatographs can separate very complex mixtures of very minute samples of fatty acids, and so provide an extremely useful tool. The method can of course be used to separate mixtures of many other compounds in biochemistry.

Soaps

Hydrolysis of triglycerides was mentioned above as a method of breaking these compounds down into their constituent parts, glycerol and the fatty acids, the fatty acids being produced as salts if basic hydrolysis is used. Basic hydrolysis in fact forms the basis of one of the most important reactions of the triglycerides, called saponification, for the sodium and the potassium salts of the long, straight chained fatty acids are called soaps. Ordinary commercial soap, the stuff little boys claim to hate, is basically sodium stearate, and is thus manufactured by the saponification of beef tallow by sodium hydroxide solution.

Soaps owe their cleansing and grease dissolving properties to their molecular structure. Hydrocarbons with long carbon chains are insoluble in water and do not dissociate into ions. They are thus called non-polar molecules, and so the hydrocarbon chains of the soap molecules are called the non-polar regions. The 'salt end' of the soap molecules, however, does dissociate into ions, giving rise to a positively charged sodium ion. This region is therefore described as polar, and because it dissolves in water, the whole soap molecule is able to dissolve with it. When soap meets grease, for instance on a shirt, the hydrocarbon non-polar region of the soap molecule dissolves in the grease, but because the polar end of the soap is still free, the whole comple͙ ˙ can dissolve, grease and all, and leave the shirt spotless. The polar and non-polar aspects of molecules are very important biochemically, and we shall meet this phenomenon again when discussing some of the compound lipids.

It is rather interesting that soap production was known to man as far back as the days of the early Babylonians. These people used water extracts of wood ash instead of sodium hydroxide in their saponification process, as this extract contains sodium and potassium carbonates, which act as mild bases. This process was used right up until the twentieth century, when sodium hydroxide became available widely, and in fact it is still used in some primitive countries even today.

Another property of fats which we meet every day is their oxidation. When exposed to the air, a saturated fat is oxidized, and some of the single bonds are dehydrogenated to double bonds, and the fat hence becomes liquid as its melting point is lowered by the presence of these double bonds. This liquefaction is the first stage in the process whereby butter goes rancid, and the process is continued by slight hydrolysis of some of the esters giving rise to free fatty acids, among them the evil-smelling butyric acid which gives rancid butter its characteristic odour.

Fats and oils in living organisms

In living organisms, fats and oils have two main functions: as reserves of food material and as structural components of living tissues. As a food material, fat is very valuable. Although not so widely used as glucose and the other monosaccharides, it has about twice the calorific value of these carbohydrates—when the animal uses fat as food it obtains much more energy from it than it could from sugars; in fact fat is the most energy-rich food we know. But more interesting than its immediate food value, is that fat provides a reserve of food materials for times of need. Foods which are ingested in excess of immediate requirements—and this includes carbohydrates and proteins as well as fats themselves—are converted to fat by the organism and deposited in certain areas of the organism's body. When food materials are scarce, the organism can convert some of this store to immediately usable forms and hence survive. In mammals some 10 % of the body weight may be in the form of fat, to the horror of some humans, but to the distinct advantage of those mammals which hibernate, as they are thus able to survive the whole winter without eating. In plants, fat reserves are found in abundance only in seeds, where they form food reserves for the embryonic plant during germination and the early stages of growth.

Sometimes the fat reserves of animals can have other functions. In the main they seem to occur around important and delicate organs such as the kidney and the liver, where they can provide a cushion for these organs against sudden shocks. But it is not known whether this function is by design or merely an accident of evolution.

The structural functions of the simple fats are few, but vital. They are found in many tissues, but especially in biological membranes, for instance the plasma membrane which segregates the cell from its environment, and generally the fats provide a bridge for the gap between water soluble and water insoluble phases. In this function their polar and non-polar character is of importance, and they are assisted in this role by some of the compound lipids which will be discussed later.

The waxes

The remaining members of the simple lipids are the waxes, materials we met at the beginning of this chapter as the candles and the coatings on the outsides of leaves. The waxes are also esters, although the alcohol involved is not glycerol but a monohydric alcohol with a very long carbon chain. Beeswax for example is mainly an ester of palmitic acid with the straight chain alcohol myricyl alcohol, which has the formula $CH_3(CH_2)_{29}-OH$, that is a hydrocarbon chain 30 atoms long with a primary hydroxyl group at the end. The cutical waxes which form the coatings of flower petals, fruit skins and leaves are composed of long fatty acids with chain lengths of between 24 and 35 carbon atoms, linked as esters with very long chain primary and secondary alcohols.

Waxes may sometimes act as food stores in the same way as the fats, but more often their role is protective, as they can form a surface layer which is repellent not only to water but to insects and bacteria, providing a barrier against disease.

Compound lipids—the phospholipids

The compound lipids, as might be expected from their name, have a more complex structure than the fats, the oils and the waxes. Apart from an alcohol and a mixture of fatty acids, these compounds contain other groups such as phosphoric acid groups, nitrogen containing bases, carbohydrates and proteins. But we shall first consider the group, based on glycerol, called the phospholipids.

The common phospholipids include several derivatives of a compound called phosphatidic acid, where a phosphoric acid group is joined to one of the hydroxyl groups of a glycerol ester through an ester linkage. Phosphatidic acid, although rare in its free form in animals, is common in many plants, such as cabbage. Because this molecule forms the basis of this series of compounds, they are sometimes referred to as the phosphatides. The addition of a nitrogenous base molecule to the phosphoryl group now transforms this molecule into a phospholipid, and we need be concerned with three different compounds here, those whose structures are shown in diagram 21.

If the nitrogenous base is choline, the lipids are called phosphatidyl cholines, or referred to by their common name, lecithins.

If the base is ethanolamine, the lipid is called phosphatydyl ethanolamine.

If the base is serine, the lipid is called phosphatydyl serine.

Together the phosphatidyl ethanolamines and the phosphatidyl serines are called the cephalins. The different compounds in each of these groups are a consequence of the different hydrocarbon chains attached to the glycerol molecule, and they are generally derived from the same long chain fatty acids which are common in the simple lipids. As a general rule in the lecithins, the hydrocarbon chain next to the phosphoric acid group is unsaturated and could be oleic or linolenic, while that on the other terminal carbon atom of the glycerol is saturated, either palmitic or stearic acid.

The phospholipids, unlike the simple lipids, are not concerned with food storage at all, but are intimately involved in biological structures. Brain and nerve tissue contains a very large proportion of these

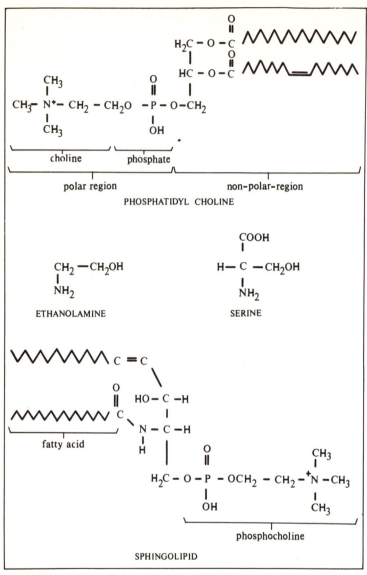

21. *Structures and components of some compound lipids.* Phosphatidyl choline, a phospholipid, is composed of glycerol, two fatty acids, one of which is usually unsaturated, a phosphate group and choline. Other phospholipids contain ethanolamine or serine in place of choline, and they all have a polar and a non-polar region. Sphingolipids are composed of a fatty acid region, a phosphocholine structure, and the alcohol sphingosine, which comprises all the rest of the molecule which is not labelled

materials; some 25 % to 30 % of the dry weight of brain tissue may be made up of phospholipids, compared with only half this proportion in other tissues such as the liver.

Lipids in biological membranes

In many ways, the role of the phospholipids in the biological membranes is the most interesting. Examination of the structure of phospholipids shows that they all have a non-polar region occupied by the long hydrocarbon chains, and a very polar region around the phosphoric acid and nitrogenous base areas, so these molecules have a number of the same characteristics as the soaps as far as structure is concerned. The polar regions are thus able to dissolve in aqueous (watery) phases. Biological membranes such as those which surround the individual cells of living organisms, and which form the basic structures of all the various minute structures within the cells—the organelles concerned with the production of energy, the synthesis of organic materials such as proteins, and the protection of the cell from invasion by harmful substances and organisms by the production of corrosive chemicals which attack and destroy them—all are composed of layers of lipids and layers of proteins. Electron microscope pictures of all these membranes show that they are composed of two dense layers corresponding to protein material, enclosing a lighter inner zone which is largely made up of lipid. The very central core of these membranes is usually simple fatty material, and so we have the situation where there is an interface between fat and water, as proteins are very polar substances which dissolve in water.

The biochemical role of the phospholipids in membranes is to bridge this gap between the polar and the non-polar regions. Thus the long hydrocarbon region of the molecule is embedded—one could even think of it as being dissolved—in the fatty region in the centre of the membrane, while the other end of the molecule, the site of the phosphate group and the nitrogenous base, is intimately associated with the aqueous region on either side of the membrane structure, the region which is made up largely of proteins. And in this way the whole phospholipid molecule forms a link between the inner fatty core and the layers of protein on either side, 'tying' them together and so performing a vital role in the stability of these extremely important biological structures.

Sphingolipids

The next important group of compound lipids are the sphingolipids. These compounds differ widely from the fats and the oils, and even from the phospholipids, in that they are not based on the trihydric alcohol glycerol. In fact they are not based on a trihydric alcohol at all, rather a dihydric alcohol with an amine group attached to its carbon chain, in other words an amino-dialcohol, called sphingosine. This is a long chain molecule, made up of 18 carbon atoms with a double bond on which the hydrogen atoms are in the trans configuration. The hydroxyl groups are attached to the terminal carbon atom (carbon number one) and carbon number three, while the amine group is sited between them on carbon number two. The major sphingolipids are called the sphingomyelins, taking their name from their function in living tissues of forming part of the myelin sheath, the coat which is found around nerve fibres; an example is shown in diagram 21 on page 76.

It will be noticed immediately that there are further differences from the phospholipids, as well as the involvement of a different alcohol. The fatty acid is joined not to a hydroxyl group, but to the amine group, forming a type of linkage which we have not met before, an amide linkage. This linkage is formed between a carboxyl group and an amine group, the hydroxyl of the carboxyl group and one of the hydrogen atoms of the amine group being driven out as a molecule of water, leaving the two molecules joined together, so the mechanism of this reaction is not very different from the ordinary ester linkage formation. The terminal hydroxyl group of the sphingosine is linked to a phosphoric acid group, which in turn is linked to a choline molecule, the same compound involved in the lecithins.

Other complex lipids

Finally, we come to a group of the lipids which are associated with other complex biochemical compounds, the lipoproteins and the glycolipids. In the lipoproteins, lipids are associated with proteins; they are often compound lipids such as the phospholipids, and these complexes are seen predominantly in the membranes of living tissues. But it is reasonably certain that these complexes are merely loose combinations of discrete lipid and protein molecules which are held together only by physical forces, and there is probably little or no convalent bonding between the two types of compound, so we need trouble no further with these materials.

The glycolipids on the other hand are a special class of compounds where lipid and carbohydrate material is linked convalently into complex molecules, and these chemicals are found associated with phospholipids in many tissues, especially in the brain. They are, however, quite different from the phospholipids and from other compound lipids which we have been looking at in that they contain no phosphorus in their structures. As they have not yet been studied as extensively as other lipids or other biochemical compounds, we can only describe their structures in very general terms, although two types are commonly recognized in living systems, the cerebrosides and the gangliosides.

Cerebrosides appear at first glance to be structurally related to the sphingomyelins, but the phosphoric acid group and the nitrogenous base choline are replaced in these compounds by the molecule of a monosaccharide—usually galactose, rarely glucose—in a glycoside linkage with the hydroxyl group on carbon atom number one of the sphingosine molecule. The fatty acid joined in amide linkage with the amine group on carbon two of the alcohol has 24 carbon atoms in its chain, and has either a double bond between carbon atoms two and three (counting the carbonyl carbon as number one) or a hydroxyl group on carbon number two. The gangliosides have structures very similar to those of the cerebrosides—sphingosine, a fatty acid and a carbohydrate—but the carbohydrates involved are generally much more complex; they are often aminosugars or a monosaccharide called neuramic acid. The different gangliosides differ in the number and kind of sugar molecules found in the molecules.

The steroids

The compounds known generally as the steroids are normally included in any study of the lipids, although, as can be seen immediately from the structure in diagram 22 on page 81, they bear absolutely no structural relation to them. The steroids include a number of extremely important compounds, among them many of the substances called hormones which control the functioning of specific organs and parts of living organisms, and they are also involved, along with the fats and the compound lipids, in the structure of biological membranes. Combined with monosaccharide molecules, they form a group of materials with important properties as drugs and poisons.

The common characteristic of the steroids is the possession of a complex carbon ring system which delights in the name perhydro-cyclopentanophenanthrene ring. Although this name tells the biochemist all he needs to know about the structure of this ring—that it is composed of four rings, three of which are six-membered and one five-membered—it is much simpler, and less tongue twisting, to use its common name, the sterane ring. In the parent compound, sterane itself, each carbon atom is bonded to its maximum number of hydrogen atoms, and is therefore fully saturated. Different steroids are recognized by the different groups and atoms which are attached to this basic structure in different positions, and by the length of the carbon chain, called the 'side chain', attached to carbon number 17. Usually there are two methyl groups ($-CH_3$) attached to the sterane ring system, joined to carbon numbers 10 and 13, and the carbon atoms of these methyl groups are numbered 19 and 18 respectively. The carbon atoms in the side chain are thus numbered from 20 onwards, and commonly there are between 2 and 11 carbons in this chain. Of the other possible functional groups, the most important are the hydroxyl groups and the carbonyl groups joined to carbon atoms which form the rings. There is further variety in that double bonds can be formed either within the ring or in the side chains.

The best known steroid is cholesterol, familiar from those advertisements which urge us to eat margarine because it contains no cholesterol. This is because margarines are often prepared by 'hardening' vegetable oils which do not contain cholesterol, while butter, of animal origin, contains a great deal of this substance; we are advised to eat less cholesterol because an excess can have harmful effects on the body.

Cholesterol is a member of a class of the steroids called the sterols, molecules which have a hydroxyl group on carbon number three in the ring system, and a side chain of between 8 and 10 carbon atoms. Cholesterol itself contains 27 carbon atoms altogether, so that it has a side chain of 8 carbon atoms. It plays a structural role in the membranes, where, associated with the phospholipids, it seems to interact with them in some way and to endow the membrane with a greater degree of stability than would be possible with just phos-

22. Common steroid structures are based on a ring system of four rings of carbon atoms. Cholesterol, with a side chain of eight carbon atoms, is the most important of the sterols, steroids with a hydroxyl group in position 3. The two other structures are steroid hormones, testosterone being the male hormone and oesterone the female hormone. In the body they are responsible for the normal development of the secondary sexual characteristics

pholipid alone. It also provides the starting point from which living systems manufacture a number of the important hormones. Biochemists who wish to study cholesterol use as their laboratory tissues blood, milk and egg yolk, as these materials have the greatest concentrations of the compound.

The principal hormones based on the sterane structure are the sex hormones, responsible in the body for the development of the secondary sexual characteristics such as hair distribution, pitch of the voice, and the mammary glands. These hormones can have molecules with 21 carbon atoms, that is with just a short side chain, 19 carbon atoms with no side chains at all, or 18 carbon atoms, with no side chain and even lacking the methyl group on carbon atom number 10.

The structures of the male hormone testosterone and the female hormone oesterone are shown in diagram 22, and it is a sobering thought that the principal difference between these hormones, and thus by extension the differences between a man and a woman, is just one methyl group and the relative positions of a carbonyl group and a hydroxyl group. It should be noticed, however, that the 18 carbon hormones have one of their rings now containing double bonds, just like the benzene ring we met much earlier. Thus the hydroxyl group in this molecule has the characteristics of a phenol group rather than of a simple alcohol as do the other hormones.

Steroid glycosides

For a last look at the variety of the steroids, it is interesting to digress a little from pure biochemistry into the realms of animal camouflage. Certain complex steroids are found in nature joined to carbohydrates by glycoside linkages, and are called steroid glycosides. They are valuable as drugs used in treating heart conditions, since they can influence the action of the heart, but in any more than minute doses they are quite poisonous to animals. Plants such as the common milkweed produce large quantities of these steroid glycosides, and it is a theory that this protects them from attack by plant-eating animals. A species of butterfly has found that it too can use these compounds to advantage. It lives exclusively on the juices of these plants and accumulates the toxic steroid glycosides in its body. From past experience, birds which normally live on insects have learned to give wide berth to this type of butterfly, and they probably use its very characteristic colourings and markings to identify and steer away from danger. Here we have an instance where the external appearance of an animal can tell us something, albeit not a great deal, of the animal's molecular make-up. Unfortunately, the biochemist cannot rely on external appearance to tell him of the complexities of the biochemical compounds hidden within living systems, and has to resort to much more intricate methods of investigation. We shall meet some more of these in the next chapter, which deals with perhaps the most important of the biochemicals, the proteins.

5

The Amino Acids and Proteins

Probably the most important, and certainly the most versatile, of all the chemical constituents of living matter are the proteins. They form a large proportion of the structures of living organisms (in man, proteins form about half the total dry weight of the body) and there is hardly a living tissue or biological activity in which they are not involved in some major way. It has often been said of proteins that, biologically speaking, they can do anything, and for this reason alone they are the most interesting of all biochemicals to study.

The major protein groups

The number of possible different proteins is astronomical, for reasons which will become clear later, but they can be thought of in general terms as being divided into two major groups. By far the most abundant are the proteins whose molecules are long straight fibrous structures, and these are involved as structural materials. They are the chief components of muscle fibres, of skin, of nerves and of tendons, which is why we regard meat as an important 'high protein' food. A protein called collagen, found in cartilage, is also involved in the stability of the skeleton of animals. Hair and finger-nails are also almost completely protein material, with long fibrous structures.

The other group of proteins have molecules which are folded into almost spherical shapes; these are called the globular proteins. Although less abundant than the fibrous proteins, they are, if anything, even more important, for they control the everyday workings of the chemical machine of life. Antibodies, the substances which help to protect us against disease by seeking out and inactivating foreign and harmful agents such as bacteria, are made of proteins. Certain hormones are also proteins, and, just as with the steroid hormones which we met in the last chapter, these substances help to control the smooth running of the living system and the functioning of special parts and organs.

The most important of all the proteins are the globular proteins which we call the enzymes. All the complex biochemical reactions which we associate with life would, on their own, take place extremely slowly, in fact sometimes so slowly that one could not detect that any reaction had taken place even after periods of months or years. Obviously life based on such slow reactions could not possibly exist; if an organism had to wait six months for a molecule of glucose to be broken down to release its energy, it would be in a very sorry state. When a chemist wants a particular reaction he is working with to proceed faster than it would under normal conditions, he adds what he calls a catalyst, a substance which will accelerate the reaction, but will not itself take part or be changed by it. To take a simple, non-chemical, example, if a horse will not go fast enough, the rider may strike it with a whip. This produces the desired reaction, that is, a speeding up of the horse, but the whip, the catalyst, remains unchanged, and can be used again at another time or on another horse to produce the same effect.

Enzymes are biological catalysts. Living systems use them to promote or to accelerate biochemical reactions, but they do not themselves become changed by the reaction. Without enzymes, life would grind to a halt; they catalyse all the reactions in which the organism manufactures the chemicals and structures of life. Not least among these reactions are those in which the enzymes are made.

The very name 'protein' acknowledges their supreme importance. It was coined as far back as 1838 by a Dutch scientist G. J. Mulder from the Greek *proteios*, which means 'of the first rank', and he so named these materials because he was the first to recognize their vital role in life. Since that time, they have held the interest of a great many scientists, and today we know a great deal of their intricate structures and their functions. Even so, we find that as we learn more of the proteins, and in fact of any of the chemicals of life, new fields of research lie before us. The more we learn, the more we see there is to learn, and really we have only just started.

The amino acids

Proteins, like the polysaccharides, are composed of long chains of smaller, simpler molecules joined together. The building blocks of the proteins are called the amino acids, and before we can go on to look at the proteins themselves, we must examine these simpler materials and learn of their structures and their functions. The amino acids contain the elements carbon, hydrogen, oxygen, nitrogen, and sometimes sulphur, and they are all, in their pure states, whitish powders with faint but distinctive smells. They are generally soluble in water, but insoluble in organic solvents such as ether, and have high melting points. Usually melting is accompanied by decomposition of the amino acids.

Amino acids, as the name indicates, are organic molecules with two important functional groups, an acidic carboxyl group and a basic amine group. Any organic molecule which contains these two groups can rightly be called an amino acid, but in the present context of proteins, we are concerned with only a special few of them. These have their amine group in a position α to the carbon atom of the carboxyl group: that is they have their amine group attached to the carbon atom which is next to the carboxyl group in the molecule. The general structural formula of the amino acids is shown in diagram 24 on page 94, and it can be seen that another way of describing these is to say that both the carboxyl group and the amine group are attached to the same carbon atom. But, whatever way one looks at it, these special amino acids are called the α amino acids, because of the position of the amine group. The identity of the R group on the carbon atom determines the identity and the properties of the particular amino acid, and it is on the basis of this group, normally called the side chain, that the protein amino acids are classified. This will be discussed in a later section on the individual amino acids.

The discovery of the amino acids dates back to 1806, when a compound was isolated from the juice of asparagus. It was called, appropriately, asparagine, but little was discovered about this compound at the time. The first amino acid to be isolated from protein material was obtained in 1820. When the protein gelatine was broken down, and the mixture of products separated, it was found that the most abundant was an amino acid. This was named glycine, because of its sweet taste. It is rather a coincidence that glycine, the first amino acid to be isolated in crystalline form from protein material, is also the simplest of all the amino acids found in proteins.

Between 1820 and 1935, 18 more amino acids were isolated from proteins, making a total of 20 altogether. As they were discovered, they were given names in an arbitrary way, names which had little or no relation to their chemical structures or properties, but while other common, or trivial, names have largely been superseded in other branches of chemistry and biochemistry in favour of systematic names which described the compounds chemically, the common names of the amino acids are still with us, and are used much more frequently than the full chemical names. After 1935 no more amino acids were found in common proteins, although a few were isolated from non-protein sources such as the cell walls of bacterial cells. These other amino acids are not so important to us in this chapter and they will only be mentioned in passing later on.

Isomerism

In the simplest amino acid, glycine, the R group is simply a hydrogen atom, but in all the other amino acids, the side chain is more complex. Thus this central carbon atom, to which the carboxyl group, the amino group, the hydrogen atom and the side chain are all attached, is 'asymmetric' because it has four different groups attached to it, and its mirror image cannot therefore be superimposed on it. All amino acids then, with the exception of glycine, show optical isomerism and can be assigned to a D or an L series, just as can the monosaccharides. The configuration of these isomers of the amino acids is referred to the configuration of D-glyceraldehyde in an attempt to assign a particular amino acid to either the D or the L series.

In order to decide which configuration was present in the amino acids, some very careful and delicate experiments had to be carried out. In fact, just one of the amino acids, serine, was used in the initial experiments, and by using chemical reactions in which the mechanism was known exactly, this compound was converted to a molecule of lactic acid which had the same absolute configuration as the serine. That is, if the serine was a member of the L series, then the lactic acid produced by these reactions would also have this configuration and be a member of an L series. The other half of this experiment was to convert, using equally strict and known conditions, D-glyceraldehyde into a lactic acid molecule. What resulted from these experiments was that when D-glyceraldehyde was changed into lactic acid, only D-lactic acid was produced, but when serine from proteins was changed into lactic acid, only L-lactic acid was produced. So it could be asserted with conviction that the serine was a member of the L series, and because in naturally occurring molecules, only the members of one of the two possible isomeric series is present, it was assumed that all amino acids present in proteins were of the L series. This fact has since been confirmed. Once again it should be remembered that assignment to this series tells us nothing of the molecule's behaviour with polarized light, and different L amino acids can rotate the plane of polarized light either to the right or to the left, the actual direction and degree of rotation depending on the nature of the side chain of the particular amino acid.

Although typical proteins are composed only of amino acids of the L series, some D amino acids do exist in nature. They are found to occur in the cell walls of some bacteria and are also manufactured by some plants, but they never occur in proteins.

A few of the common amino acids, for example threonine, have two 'asymmetrical' carbon atoms in their molecules. Remembering from the chapter on the monosaccharides that the number of possible isomers is given by the expression 2^n, where n is the number of 'asymmetrical' carbon atoms, these amino acids can therefore exist as 2^2, or 4, possible isomers. But once again nature does not seem to like having more than one optical isomer in common use, and of the four possibilities in these few amino acids, only one is involved in the structures of typical proteins.

Classification of the amino acids

A number of ways of classifying the amino acids are in common use, and they depend on the side chains and their structures. All of them are valid, but some tend to be a little laborious, with some of the amino acids falling into two classes, so we shall use just one in this book, a classification which divides the amino acids into three main groups. Those in one group have an aliphatic side chain, that is a side chain consisting of a hydrocarbon chain, with or without other functional groups attached to it. The members of the next group have a benzene ring structure in their molecules, and the third group contains more than half of all the common amino acids, and is itself sub-divided according to the functional groups attached to the aliphatic side chain. It will be useful, while reading this section, to refer to the table of the amino acids on pages 89 — 92, where the structure of each acid is given.

The first sub-division we call the monoamino-monocarboxylic

amino acids, because they contain just one of each of the important functional groups. Glycine has already been mentioned as having no side chain, only a hydrogen atom as the R group. Alanine, valine, leucine and isoleucine each have a simple hydrocarbon side chain, although in the last three, the side chains are branched. Serine and threonine also have hydroxyl groups attached to their side chains. Of the amino acids in this sub-division, both threonine and isoleucine have two 'asymmetrical' carbon atoms, giving four possible isomers, although as mentioned above, only one of each occurs naturally in proteins.

Amino acids in the next sub-division have a carboxyl group in their side chains, and so are called monoamino-dicarboxylic amino acids. In solution in water, these amino acids, aspartic acid and glutamic acid, will of course show weak acidic properties due to their extra free carboxyl group and are often called the acidic amino acids. Also included in this sub-division are the amides of these amino acids, asparagine and glutamine. Amides, it will be remembered, are, structurally, carboxylic acids in which the hydroxyl part of the carboxyl group has been replaced by an amine group, and in these amino acids it is the carboxyl group on the side chain which has been converted into an amide.

The third sub-division of the aliphatic amino acids is the basic amino acids, basic because they contain an amine functional group in their side chains; so, following the naming convention we have been using so far, they are diamino-monocarboxylic amino acids. There are just two which are found commonly in proteins, lysine and arginine.

The fourth and last sub-division is of interest because its members contain the element sulphur, and again there are only two, methionine and cysteine. The sulphydryl group, —SH, at the end of the side chain of the cysteine molecule, is able to react with its counterpart on another cysteine molecule in an oxidation reaction which results in the two hydrogen atoms being split off and the two sulphur atoms being joined in what is called a disulphide bridge. The resulting molecule is another amino acid called cystine, which, as we shall see, plays a vital role in the structure of protein molecules.

---

23. *The 20 amino acids commonly found in proteins.* Because each amino acid has an 'asymmetrical' carbon atom with four different groups attached to it, two optical isomers are possible in each case. But only those acids with absolute

configurations similar to L-glyceraldehyde, the L amino acids, are found in proteins. The table also shows the structure of cystine, which is formed from two cysteine molecules by a disulphide bridge

NAME	STRUCTURE	ABBRE-VIATION
	side chain	
MONOAMINO–MONOCARBOXYLIC		
glycine	$H-\underset{\underset{H}{|}}{\overset{\overset{NH_2}{|}}{C}}-C\overset{O}{\underset{OH}{}}$	gly
alanine	$CH_3-\underset{\underset{H}{|}}{\overset{\overset{NH_2}{|}}{C}}-C\overset{O}{\underset{OH}{}}$	ala
valine	$\underset{CH_3}{\overset{CH_3}{}}CH-\underset{\underset{H}{|}}{\overset{\overset{NH_2}{|}}{C}}-C\overset{O}{\underset{OH}{}}$	val
leucine	$\underset{CH_3}{\overset{CH_3}{}}CH-CH_2-\underset{\underset{H}{|}}{\overset{\overset{NH_2}{|}}{C}}-C\overset{O}{\underset{OH}{}}$	leu
isoleucine	$CH_3-CH_2-\underset{\underset{CH_3}{|}}{CH}-\underset{\underset{H}{|}}{\overset{\overset{NH_2}{|}}{C}}-C\overset{O}{\underset{OH}{}}$	ile
serine	$OH-CH_2-\underset{\underset{H}{|}}{\overset{\overset{NH_2}{|}}{C}}-C\overset{O}{\underset{OH}{}}$	ser
threonine	$CH_3-\underset{\underset{OH}{|}}{\overset{\overset{H}{|}}{C}}-\underset{\underset{H}{|}}{\overset{\overset{NH_2}{|}}{C}}-C\overset{O}{\underset{OH}{}}$	thr

NAME	STRUCTURE	ABBRE-VIATION
	side chain	
MONOAMINO – DICARBOXYLIC (ACIDIC)		
aspartic acid	$\overset{O}{\underset{HO}{\|\|}}C - CH_2 + \overset{NH_2}{\underset{H}{\|}}C - C\overset{O}{\underset{OH}{\|\|}}$	asp
glutamic acid	$\overset{O}{\underset{HO}{\|\|}}C - CH_2 - CH_2 + \overset{NH_2}{\underset{H}{\|}}C - C\overset{O}{\underset{OH}{\|\|}}$	glu
asparagine	$\overset{O}{\underset{NH_2}{\|\|}}C - CH_2 + \overset{NH_2}{\underset{H}{\|}}C - C\overset{O}{\underset{OH}{\|\|}}$	asn
glutamine	$\overset{O}{\underset{NH_2}{\|\|}}C - CH_2 - CH_2 + \overset{NH_2}{\underset{H}{\|}}C - C\overset{O}{\underset{OH}{\|\|}}$	gln
DIAMINO – MONOCARBOXYLIC (BÁSIC)		
lysine	$NH_2 - CH_2 - CH_2 - CH_2 - CH_2 + \overset{NH_2}{\underset{H}{\|}}C - C\overset{O}{\underset{OH}{\|\|}}$	lys
arginine	$NH_2 - \overset{}{\underset{NH}{\overset{\|}{C}}} - CH_2 - CH_2 - CH_2 + \overset{NH_2}{\underset{H}{\|}}C - C\overset{O}{\underset{OH}{\|\|}}$	arg

NAME	STRUCTURE	ABBRE-VIATION
	side chain	
SULPHUR CONTAINING		
methionine	CH_3— S —CH_2—CH_2 ┤ C — C (NH$_2$, O, OH) ├ H	met
cysteine	HS—CH_2 ┤ C — C (NH$_2$, O, OH) ├ H	cys
cystine	CH_2 ┤ C — C (NH$_2$, O, OH) ├ H S S CH_2 ┤ C — C (NH$_2$, O, OH) ├ H	cys
AROMATIC		
phenylalanine	CH_2 ┤ C — C (NH$_2$, O, OH) ├ H	phe
tyrosine	HO— —CH_2 ┤ C — C (NH$_2$, O, OH) ├ H	tyr

NAME	STRUCTURE	ABBRE- VIATION
HETEROCYCLIC	side chain	
tryptophan		try
histidine		his
IMINO		
proline		pro

The other two groups of the amino acids can be discussed together. Phenylalanine and tyrosine both contain an aromatic ring structure in their molecules, tyrosine in addition having a hydroxyl group attached to the aromatic ring, so that this part of the side chain really resembles a phenol group. This amino acid, together with one of the heterocyclic amino acids, trypotophan, has the property that it is able to absorb ultraviolet radiation, so that if ultraviolet radiation is passed through a sample of a protein and found to be absorbed intensely, it can be deduced that the protein has some of these molecules in its structure. The other true heterocyclic amino acid is histidine. The final amino acid, proline is really a cyclic amino acid, the amine group having become part of a ring, and so it has a secondary amine group instead of the more usual primary amine groups contained by all the other amino acids. The molecule is sometimes called an imino acid, an old and somewhat confusing name which has unfortunately stuck.

This concludes the classification of the amino acids on the basis of their side-chains, but there is another form of classification which is sometimes used. This refers not to the structures of the molecules, but to their role in nutrition. Although living organisms can synthesize amino acids for themselves from simple raw materials, each animal or plant requires a number of amino acids which it is unable to make for itself. The animal or plant must therefore obtain these amino acids intact from its food and make direct use of them for making proteins, and these amino acids are called the essential amino acids for that organism. Man, for example, is unable to make valine, lysine, threonine, leucine, isoleucine, tryptophan, phenylalanine or methionine for himself and so he must ensure that he eats food materials which contain these vital molecules. Most animal-derived food, such as meat, fish, eggs and cheese contains proteins which are made up from all of the amino acids we have been considering, and therefore these proteins are called complete proteins. But vegetables and other forms of plant foods contain few complete proteins; usually they lack lysine, threonine and trypotophan, and so for good health and strength, man must include some animal products in his diet. Pure vegetarians, who will not eat eggs or milk, put themselves at a distinct disadvantage for this very reason.

Properties of the amino acids

Properties of amino acids are dictated both by the side chains and by the amine and the carboxyl functional groups. Side chains may be almost inert, such as the hydrocarbon side chains of the first sub-division of group one, or they may be highly reactive. We have already seen that the sulphydryl groups of cysteine can become involved in disulphide bridges, that the extra carboxyl groups on aspartic and glutamic acids make these molecules highly acidic and thus able to react with bases to form salts, and that lysine and arginine owe their basic properties to the presence of the second amine group, giving them also the ability to form salts. These properties will become more important when the amino acids are joined together in long protein chains, but for the present we can concentrate on the properties of the amine and carboxyl groups present in every amino acid on the same carbon atom.

Zwitterions

Some of the properties of the amino acids are quite different from those which would be expected by looking at their structures. With few exceptions they are soluble in water, but insoluble in ether, acetone and other organic solvents, although just the opposite would be expected, as organic amines and carboxylic acids are generally soluble in organic solvents but insoluble in water. Melting points of organic amines and carboxylic acids are usually low, but the amino acids only melt at high temperatures and the molecules are often decomposed on melting. Overall, the structures and functional groups of the amino acids would suggest that they are typical organic compounds which do not show polar properties and do not dissociate into ions, while these strange observed properties indicate that they are in fact charged, polar compounds.

More evidence in support of the theory of the polar character of the amino acids is provided by examining their behaviour when dissolved in water. If a pair of wire electrodes connected to a battery is immersed in an amino acid solution, the amino acid molecules will not move towards either electrode. If the pH of the solution is raised by the addition of a base, such as sodium hydroxide, the molecules begin to move towards the positive electrode, or the anode, suggesting that they have become negatively charged anions. If on the other hand acid is added to the solution, so that the pH is lowered, the

24. In solution in water amino acids can exist in a number of ionized forms. If both the amine and the carboxyl groups are ionized, the amino acid is said to be in zwitterion form, and there is no net electric charge on the molecule. Addition of either base or acid neutralizes one of the ions, so that a net charge is left, and the amino acid will migrate towards one pole of a battery when an electric current is passed through the solution

molecules will behave as positively charged cations, and will move towards the negative electrode, the cathode. In other words, amino acids in solution in water will react with both acids and bases, and because of this dual behaviour they are described as amphoteric compounds—having both acidic and basic properties.

The amphoteric nature of amino acids is the result of the ionization of both the functional groups in the molecules. In neutral solution in water, both the amine and the carboxyl groups are in their fully charged form, the carboxyl group having given up a proton by dissociation and thus acquired a negative charge, and the amine group having accepted a proton and become positively charged. In this form the molecules are called zwitterions, a name coined by Bjerrum in 1923 from the German for 'ions of both kinds'.

When a base is added to a neutral solution, the proton from the positively charged amine ion is removed and reacts with hydroxyl ions to form water, leaving the amine group uncharged. Thus the only charge left on the amino acid molecule is the negative charge of the carboxyl ion, so that the whole molecule carries a net negative charge and will migrate towards the positively charged electrode. Conversely, when acid is added to a neutral solution, lowering the pH, protons from the acid will react with the carboxyl ion, neutralizing it to an uncharged carboxyl group. Now the only charge left on the amino acid molecule is the positive charge on the amine ion, and the whole molecule will migrate towards the negatively charged electrode. These events are summarized in diagram 24.

Other reactions of the amino acids

Two rather special reactions of the amino acids are of interest to us here, and will be met later on in this chapter. The amine group of the amino acid can react with a compound called 2.4.dinitrofluorobenzene to give a 2.4.dinitrophenyl derivative of the amino acid, which is a yellow coloured compound. The other reaction is that of amino acids with a chemical called ninhydrin, which produces a blueish-purple coloured compound. It is not important for our present purpose to know the mechanisms of these reactions or the structures of the 2.4.dinitrofluorobenzene or the ninhydrin, but just to know that the reactions take place producing coloured compounds which can demonstrate the presence of the amino acids.

Most of the reactions of the amino acids have now been mentioned, at least in principle. The carboxyl groups can react to form the usual derivatives of the carboxylic acids, such as esters and salts, but the most important is the reaction with amines to form acid amides. The amine groups also participate in the usual amine reactions, including salt formation and of course also react with carboxylic acids to form amides. The importance of the amide reaction now becomes clear, for if the carboxyl group of one amino acid undergoes an amide forming reaction with the amine group of another amino acid, we have a method by which two amino acids can be joined together through the amide linkage. And because one of the two amino acids still has a free amine group and the other has a free carboxyl group, other amino acids can become joined on to this pair, to make a long chain of amino acids. As stated early in this chapter, long chains of amino acids form the proteins.

The proteins

The amide formed when two amino acids join together is more usually called a peptide, and hence the amide linkage between the two is called a peptide bond. As each peptide bond is formed, a molecule of water is eliminated from the two molecules, a hydroxyl group from the carboxyl group of one amino acid combining with a hydrogen atom from the amine group of the other. Proteins, then, are long chains of amino acids joined together by peptide bonds. As with the polysaccharides, there is no standard definition of the number of amino acids which generally constitute a protein. A few amino acids joined together, usually up to about 10, form chains called oligopeptides. More than 10, and the chain is called a polypeptide, and if the chain contains more than about 70 (some say more than 100) amino acids molecules, it is called protein. But 70 amino acids make a very small protein, and some proteins contain many hundreds or thousands of amino acids in their chains, with molecular weights of several million. The only other requirement before long amino acid chains are called proteins is that they must be made up of the L forms only of the molecules; if some D forms are present, the chain is not strictly a protein.

The structures of protein molecules are much more complex than those of any of the other biochemicals we have discussed so far, so complex in fact that we recognize four different levels of structural organization, called the primary, the secondary, the tertiary and the quaternary structures. These levels of organization will now be described in some detail.

Primary structure of proteins

To state the primary structure of a protein chain is to give a description of the sequence of the amino acids in the chain. This may seem a fairly simple problem. We remember the section on the polysaccharides, where the aim was to find out which monosaccharide units were present in the long chains, and in what order. But in the polysaccharides, only two or three different monosaccharides are commonly encountered, and in the majority of cases the molecules are made up of repeating sequences of disaccharides. In the proteins, there are about twenty different amino acid building blocks and there are no repeating sequences of two or three units. Furthermore the order of the amino acids in the protein chain is of vital importance. For example, the two amino acids glycine and alanine can be joined together in two different ways to compose a dipeptide: either the amine group of the glycine can be joined in a peptide bond to the carboxyl group of the alanine, or vice versa, and the two possibilities form completely different compounds. If three different amino acids are joined together, there are six different possible arrangements, and again all produce different compounds. The more amino acids involved, the greater the number of possibilities. The number of pentapeptides (with five amino acids joined together) which can be made using combinations of all the twenty or so amino acids is several millions, and once again, all of them are different. By the time we come to think in terms of proteins, which have hundreds of amino acids joined together, the number of possible arrangements is beyond our comprehension. And the fact that only a minute fraction of the possibilities actually exist in nature does not help very much, for we are still faced with the formidable problem of determining the exact sequence of the amino acids in those possibilities which do occur in living systems.

Purifying proteins

The first difficulty in determining the primary structure of proteins is in obtaining pure samples of the proteins. Trying to determine the sequence of a protein when it is all mixed up with other proteins in a complex mixture is about as likely to succeed as trying to find a person's telephone number in a London telephone directory which, instead of being in alphabetical order, is arranged in a random fashion, when you do not even know the person's surname or address, just his first initial. The immensity of the problem is further amplified by the fact that it is still difficult to decide whether a protein is pure or not. As the years go by, the criteria of purity change, and what was thought to be a pure protein a few years ago now turns out to be a mixture of three or four proteins.

There are, however, a number of ways of purifying proteins and ways in which some of them can be tested for purity. The first method to be used for obtaining pure proteins was that of fractional solubility. Some proteins dissolve to a different extent from others in certain salt solutions or organic solvents; so a protein mixture can be dissolved in, say, water, and then a salt added so that those proteins which are insoluble in the salt solution are precipitated. As a rough method of separation, this method is still used today, but for any reasonable degree of purity in the final product, it is of little value. Another method is to spin the mixtures of proteins in high speed centrifuges, so that the different proteins are separated according to their different weights. This is a better method than fractional solubility, but still does not give sufficient purity in the final product.

The two most useful present day methods for separating protein mixtures are electrophoresis and chromatography. Chromatography was mentioned in the last chapter as a separating technique; by carefully selecting the stationary and the mobile phases for each mixture and for each protein to be separated, it is possible to obtain a reasonably pure compound. The process can be repeated several times with each spot or band obtained, so that the protein becomes purer with each separation.

Electrophoresis is a technique which depends on the electrical charges carried by the protein molecules, and is really a very sophisticated form of electrolysis. The amino acids in the protein chains do not have their amine and carboxyl groups free, as these are involved in the peptide bonds, so that the behaviour of proteins as electrolytes depends on the ionizable side chains. For example the diamino-monocarboxylic amino acids have amine groups in their side chains which are still free in the protein, and in solution these amine groups are present as positively charged ions, having accepted

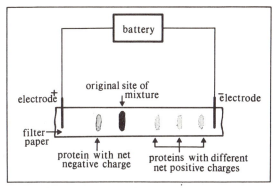

25. Proteins carry a net electric charge due to the ionizations of the side chains of their constituent amino acids. A mixture of proteins with different net charges can thus be separated by passing an electric current through the mixture in the method of electrophoresis. Placed on a strip of filter paper soaked in a suitable solvent, the components of the mixture will migrate towards one of the electrodes depending on their charges, those with higher charges moving faster than those with lower charges, so that eventually each protein is separated into a distinct band

a proton. In a similar way, the monoamino-dicarboxylic amino acids have a carboxyl group in their side chains, and in solution these lose a proton so that they carry a negative charge. In a long protein chain there will be many negatively charged side chains and many positively charged ones, and so the net charge on the whole protein molecule will be the difference between the numbers of each. If a protein has, say, 15 glutamic acid molecules in its chain, and 25 lysine molecules, it will have a net positive charge of 10, the difference between the numbers of acidic and basic side chains. Because each protein has different numbers of each amino acid in its chain, the net charges on each member of a group of proteins will be different. If an electric current is passed through a mixture of these proteins, each protein will move towards an electrode depending on its net electric charge: the higher the net charge the faster it will migrate.

This forms the basis of the technique of electrophoresis. In practice a sample of a mixture of proteins is placed on a support, such as a strip of filter paper, and the support soaked in a suitable solvent. The poles of a battery are placed on each end of the support so that a current flows along it, just as a current flows through a solution during electrolysis. Each component in the protein mixture then begins to migrate towards one of the electrodes at a speed dictated by its net charge; after a time, each component protein has travelled a certain distance; no two proteins will have travelled an identical distance, and so the mixture has been separated. Both electrophoresis and chromatography are in routine use in biochemistry laboratories today, and as the techniques become more refined, better and better separations become possible.

Once a protein has been separated from a mixture, it must be tested for purity. One way is to go on and on trying to separate other compounds from it, that is to continue treating it as a mixture, and when the 'mixture' cannot be separated any more, we can be reasonably sure of its purity. As an additional check, it can be subjected to a number of different forms of separation, such as both chromatography and electrophoresis, under different conditions each time. But it is vital to take great care when trying to establish purity, as protein molecules are very delicate and can easily be damaged or broken down. Damaged proteins are not the same molecules as the proteins sought from the separation process, so if damage does occur at this stage, the object of the exercise has been defeated.

Another method of establishing purity is to investigate the functions of the proteins. Hormones and enzymes, mentioned at the beginning of this chapter, have definite functions and effects in living systems and some of these can be imitated in the test tube. For example, if the protein we are interested in is an enzyme capable of catalyzing the reaction where compound A changes to compound B, then some of compound A can be mixed in a test tube with some of the 'purified' protein enzyme, to see if compound B is produced. By carefully observing how fast compound B appears in the test tube, it is possible to determine how pure the enzyme is, for if the protein contains traces of other compounds which do not catalyze the reaction, the reaction will proceed at a slower pace than if the protein is pure enzyme.

Amino acid content of proteins

Once a pure protein has been obtained, the real task of determining its primary structure can begin. The first step might be to find out which amino acids are present, and the relative amounts of each one, and this can be done by hydrolysis of the protein. The usual method is acid hydrolysis, when all the amino acids are released intact from the protein, except for tryptophan, which is decomposed to a black insoluble residue. Hydrolysis with base will yield this amino acid intact, but has other disadvantages in that cysteine, serine, threonine and arginine are destroyed. However, using both methods independently, each amino acid present in the protein can be freed from the chain. The mixture can then be separated into the component amino acids by chromatography, and the relative proportion of each estimated. As amino acids are colourless compounds, the chromatography stationary phases must be 'developed' before they can be seen. One way to do this is to treat the stationary phase with ninhydrin, when the amino acid spots show up as blueish-purple areas. The biochemist can then write down a 'recipe' for that protein—13 valines, 37 glycines, 11 aspartic acids and so on—but the value of the list is limited, for it tells nothing of the order in which the amino acids are joined together in the intact protein. We still know little of the primary structure of the protein.

Amino acid sequence in proteins

The actual chemical methods for the determination of the sequence of amino acids in a protein have been available for years, even decades, and it was probably the enormity of the problem which discouraged scientists from using them. In some ways the methods are similar to those used to determine the structures of the polysaccharides: the chains are broken down into smaller pieces, and the sequence in each of the pieces is determined. In one way, however, the protein problem is actually simpler than the polysaccharide problem, for in the protein chains there are no branches—the amino acids are joined together in straight chains. Two or more chains can be joined together at a number of points, but not by peptide bonds, and this joining is part of a higher level of structural organization which we shall meet later.

An important part of the primary structure determination is discovering which amino acid occupies the end positions of the protein chains. Each chain has two ends which are distinguished by which functional group is free on the terminal amino acid. If the amine group is free (the carboxyl group of that amino acid being involved in the peptide bond) this is called the N terminal end (N standing for nitrogen). If on the other hand the carboxyl group is free, this marks the C terminal end (C for carboxyl). By convention, peptides are always written down with the N terminal end on the left, so that in the peptide Ala—Ser—Val—Thr—Phe, it is the alanine which has the amine group free and the phenylalanine which has a free carboxyl group. (The three letter abbreviations for the amino acids are the same as those listed in Table 23 on pages 89–92.

To determine the amino acid at the N terminal end of a protein or peptide chain, the compound is reacted with 2.4.dinitrofluoro-benzene to give the 2.4.dinitrophenyl amino acid derivative which we met when discussing the reactions of the amino acids. If the chain is then broken down by hydrolysis, the yellow 2.4.dinitrofluoro-phenyl derivative can easily be separated from the mixture by, for example, chromatography, and identified. Determination of the C terminal amino acid is slightly more difficult, but can be carried out by an enzyme which splits off the amino acids from this end one by one. The enzyme is mixed with the protein or peptide, and then the freed amino acid separated from the mixture and identified. This process has the disadvantage that as soon as one amino acid is split off from the C terminal end, the enzyme will begin to split off the next one, as this has now become the new C terminal amino acid. But usually it is possible to stop the reaction before the second amino acid is split off.

Having identified the C and the N terminal amino acids, it is still necessary to find out which amino acids lie between them, and in what order, and it is this part of the determination which is the real problem. The first primary structure of a protein was successfully worked out and published in 1956 by Frederick Sanger and his colleagues at Cambridge. He worked with the protein hormone insulin, the substance which is vital to the workings of the body because it enables the cells to make use of glucose, the substance which diabetics are unable to make. Insulin is a very small protein, really just a polypeptide, containing only 51 amino acids in its molecule, but it took Sanger ten years to work out its structure. For his efforts he was awarded the Nobel prize in 1959, by which time other scientists were well on the way to determining the primary structures of other proteins.

The basis of Sanger's pioneering method was to break down the protein chain into small fragments, determine the amino acid sequence in these fragments and then to fit the fragments together again (on paper). Fortunately there are a number of enzymes which will split protein chains at specific points, for example at the peptide bonds involving basic amino acids such as arginine or lysine, or at those involving aromatic amino acids such as tryosine and phenyl-alanine. Once the fragments have been broken off, they can be separated by the methods used to separate proteins, chromatography or electrophoresis. In this way oligopeptides with perhaps five or six amino acids can be obtained, and by using different enzymes on different samples of protein, the chains can be broken in different points in each sample so that there is overlapping of the fragments from each mixture.

The determination of the sequence of the amino acids in the fragments is, of course, much simpler than trying to work with the whole protein chain, but it is still far from easy. The usual methods used are to split off the amino acids from the ends of the fragments, one by one, with enzymes or special chemicals, until the whole sequence in the fragment is known, and the N and C terminal ends have been identified.

The final process is to fit all the pieces of the jigsaw together to get a picture of the whole protein chain. If four fragments have been obtained from a protein chain by different methods of splitting and have been found to have the sequences Phe—Val—Arg—Ser, Val—Arg—Ser—Glu—Gly, Arg—Ser—Glu and Glu—Gly—Asp—Tyr, then it is reasonably certain that somewhere in the complete protein chain is a sequence Phe—Val—Arg—Ser—Glu—Gly—Asp—Tyr. This, however, is a sequence of only eight amino acids, and so there would still be a very long way to go before the sequence of the whole protein could be established beyond doubt. It would not be definitely certain that even the four fragments above come from the same part of the chain, and this has to be established by further fragmentation of the chain in other places and by determining the sequence of the fragments. It is small wonder that it took Sanger so long to solve this mind-bending problem.

Secondary and tertiary structures of proteins

The secondary and tertiary levels of structural organization of proteins can be considered together, as they both describe the ways that the amino acid chains are folded and held together in definite shapes. Far from being a long straight molecule, proteins are coiled and bent and more than one chain may be involved. The ways in which the chains are held together is the secondary structure. Tertiary structure refers to the three-dimensional shape of the whole molecule. The relationship between the three levels of organization are shown in diagram 26 on page 106.

The most important method used in the investigations which led to the understanding of secondary structure was X-ray diffraction. In this technique X-rays are passed through compounds, and if they have an ordered structure, definite diffraction patterns of the X-rays will be produced. In crystals, for example, all the molecules are arranged in a definite pattern, and so diffraction patterns will be produced if X-rays are passed through them.

Two eminent scientists, Pauling and Corey, working in America, made a large number of studies of the X-ray diffraction pattern of small crystalline amides, and, from their results, they established a set of rules for the most stable structures of these compounds in an orderly structure. Among these rules was the condition that stable molecules must contain the maximum number of hydrogen bonds— interactions between a hydrogen atom attached to a nitrogen or oxygen atom, and another atom of nitrogen or oxygen. If models of amino acid chains are constructed according to these rules it is found that only two possibilities fulfil all the requirements. In one of them two amino acid chains lie side by side, with hydrogen bonds linking the two chains, and because of the tetrahedral orientation of the covalent bonds of the carbon atoms, these molecules have the appearance of a pleated sheet. In the other stable structure, the hydrogen bonds join together parts of the same amino acid chain, and the whole molecule looks like a helix. In fact the helix which gives the most stable arrangement is called the α helix, and in this structure each turn of the helix involves about four amino acid molecules. A simple model of this structure can be made by twisting a string of beads round a finger, the beads representing the amino acids. The turns of the helix are kept in place by hydrogen bonds between an amino acid on one turn of the helix and other amino acids on the turn below and the turn above. The structure is shown in diagram 26 on page 106, and it will accommodate any amino acid with any side chain, so that it is a general structure for all proteins.

Although hydrogen bonds are sufficient to hold the protein

molecules in their helical shape in the solid state, when the proteins are dissolved in water the molecules could uncoil and assume a random, disorientated structure. So to keep the molecules in their characteristic shapes, other forces and bonds must be involved, and the shapes which are the results of these other bonds are called the tertiary structures of the proteins. The most powerful of these other bonds is the disulphide bridge, formed between two cysteine molecules in the amino acid chain. As the cysteine molecules can be some distance apart in the primary structure of the chain, the formation of disulphide bridges means that the two distant parts of the protein chain are held close together, and that therefore a folded structure of the molecule is stabilized. Disulphide bridges can also be formed between cysteine molecules in different protein chains, thus holding the two chains together in a stable structure. In the insulin molecule, two separate amino acid chains are involved and there are two disulphide bridges holding the two chains together. There is also another bridge formed between two cysteine molecules in the same chain. Other bonds involved in tertiary structure are so-called salt bridges, formed between an acidic side chain of one amino acid and the basic side chain of another (a kind of internal salt), and simple ionic interactions, any ionized group on one amino acid being attracted by an oppositely charged side chain on another. Although the amino acid chain has a helical secondary structure, the helical chain can be folded back on itself, and the folds held in place by these strong bonds between some of the amino acids of the chain or chains.

The actual shapes of protein molecules are not nearly so well-known as the forces involved in stabilizing them, and in fact the shapes of only a few are known. This is because the shapes are so complex that to analyse them from X-ray diffraction methods would be extremely laborious, much more so in fact than the determination of the primary structure. On the basis of their general shape, however, two classes of proteins are recognized, the fibrous proteins and the globular proteins. Fibrous proteins, as their name suggests, consist of long straight protein chains, coiled into helices but generally not folded back on themselves. It is proteins of this type which make up such structures as hair and wool, muscle fibres and tendons, in fact the structural proteins of life. Because these molecules have a reasonably ordered tertiary structure, it is possible by X-ray diffraction methods to learn something of the tertiary structures.

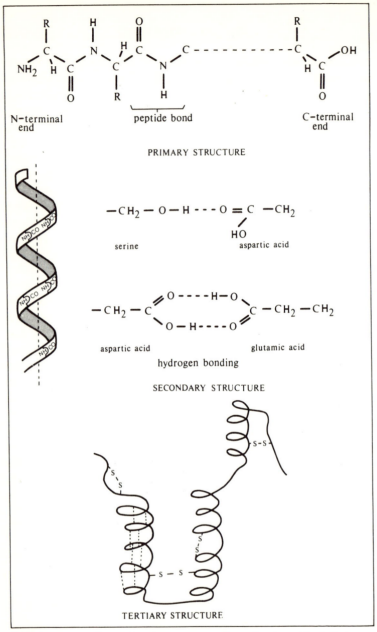

N-terminal end

peptide bond

C-terminal end

PRIMARY STRUCTURE

$-CH_2-O-H\cdots O=C-CH_2$

serine

HO

aspartic acid

aspartic acid   glutamic acid

hydrogen bonding

SECONDARY STRUCTURE

S–S

TERTIARY STRUCTURE

26. *Primary, secondary and tertiary structures of proteins.* A protein chain is composed of amino acids joined together by peptide bonds, and has a free amine group at one end (the N terminal end) and a free carboxyl group at the other (the C terminal end). The chains form themselves into a helical secondary structure, each turn of the helix involving about four amino acids, and the helix

The globular proteins, however, present much more of a problem, and it is in these molecules that the strong bonds are essential for maintaining the very complex shapes. The most obvious difference between the fibrous and the globular proteins is that the fibrous ones, which make up the structures of living systems, are largely insoluble, while the globular proteins are very soluble in water, and are sometimes referred to as the soluble proteins.

At this point it will be worth saying a little more about the enzymes, the biological catalysts. Enzymes are globular proteins, and the exact shape of their molecules is of vital importance to their function, because the shape of the molecule determines its ability to act as a catalyst. There are a number of mechanisms of catalysis by enzymes, but we shall illustrate just one as an example. See diagram 27. The reaction which is to take place is that compound A is to be combined with compound B to produce compound C. In a typical solution under normal circumstances, compounds A and B may never, or at least only very rarely, come into close contact with each other, so that a reaction between them may be very unlikely. The role of the enzyme therefore is to bring the two compounds together, and it does this by first attracting and trapping compound A on to its structure, and then attracting and trapping compound B. When the two compounds have come together on the surface of the enzyme molecule, they can join together, and the enzyme then releases them as compound C. Now, to be able to attract and hold compounds A and B, the surface of the enzyme must have special sites which are complementary in some way with each of the compounds. And these special sites are determined by the shape of the molecule, and hence by the bonds which are holding the molecule in its particular tertiary structure.

Denaturation

If the tertiary structure of a protein is destroyed, its properties change dramatically. Enzymes, for example, lose their ability to act as

is kept in place by hydrogen bonds between the amino acids of the adjacent turns. Examples of hydrogen bonding between a serine and an aspartic acid, and an aspartic acid and a glutamic acid are shown. The whole chain assumes a shape called the tertiary structure. Hydrogen bonds are shown between the coils as dotted lines, and the whole shape is further stabilized by disulphide bridges between cysteine amino acids at different parts of the chain. It should be remembered that the tertiary structure is three dimensional

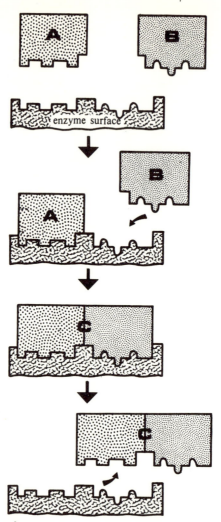

27. *Mechanism of enzyme action.* Compounds A and B are trapped and held on special sites on the surface of the enzyme molecule so that they can react and join together to form compound C, which is then released so that the enzyme can find further A and B compounds

catalysts, because their special sites are destroyed and they can no longer attract and hold the compounds while they undergo their reactions. Globular proteins which were soluble in their normal state become insoluble. An everyday example of this occurs when we boil eggs. Egg white is predominantly soluble protein, but when we heat it, it loses its tertiary structure and becomes an insoluble, hard, rubbery mass. This process of loss of tertiary structure is called denaturation, and can be caused by a number of conditions. If the pH of the solution is altered, denaturation may occur, and this can be seen by putting vinegar, a weak acid, into milk. The soluble proteins are rendered insoluble and are precipitated as an insoluble curd. Even violent shaking can denature proteins. If egg whites are beaten, they become the denatured, insoluble mass we use as meringue. This very delicate nature of proteins explains why great care must be taken when handling them in the laboratory.

Quaternary structure of proteins

Some proteins have, in addition to the three levels of structural organization we have seen so far, a fourth type of structure which describes how many different separate protein molecules are involved in a 'super-protein'. In these complexes, each protein molecule is a separate entity, although it may be composed of several amino acid chains itself. An example of a protein with a quaternary structure is haemoglobin, the red pigment of mammalian blood: in this 'super-protein' four protein molecules are involved, all held together by salt bridges. Hair fibres provide another example: each fibre consists of a number of protein molecules, all with their helical structure, but all wound round each other in a larger helix, with yet another protein molecule in the centre.

Making even a simple protein molecule, as can be well imagined, is an extremely complex task. So far man has not accomplished it, and the longest amino acid chain he has succeeded in producing has only some twenty or thirty amino acids. Yet living cells make these complex molecules with apparent ease. So the living cells must contain not only the machinery for assembling the amino acids and joining them together, but also detailed instructions for the very specific order in which the amino acids must be arranged to make a given protein. Further, these instructions must be passed on from generation to generation of cells, so that the offspring will be able to make the same proteins and to continue the process of life.

So we come to the last chapter of this book, which will describe the molecules which contain in their structures the information required to make the proteins, molecules called the nucleic acids.

6

The Nucleotides and Nucleic Acids

All the biochemical compounds we have been looking at so far in this book are familiar in some way to us in everyday life: the carbohydrates as table sugar, as starch in potatoes and as a stiffening agent in laundering shirts; the lipids as butter and cooking oils and candle waxes; and the proteins as meat and hair and fingernails. The compounds which we shall be considering in this final chapter, however, do not crop up in a recognizable form in the world around us. We cannot look around, see an object and say straight away that it is composed of nucleic acids. Nevertheless, the nucleic acids are present in every single living cell, and they have the vital role of controlling the day to day workings of the cell. In fact they form the ultimate controlling mechanisms of living systems, and can be said to hold the very secret of life.

The existence of the nucleic acids has been known for a little over a hundred years, but it is only within the past few decades that their importance has been fully recognized. This may seem surprising, but as we shall see, these compounds have extremely complex structures, much more complex even than the proteins. They are long molecules, composed of sub-units joined together, but the actual sub-units are themselves made up of smaller components, and are not simple compounds as are the monosaccharides and the amino acids which make up the carbohydrates and the proteins. Furthermore, nucleic acids are among the largest molecules known to man, some of them having molecular weights of hundreds of millions. Contained within these huge molecules are all the instructions which the cell requires to manufacture proteins, the instructions which dictate the order in which the amino acids must be arranged in the long protein molecules. And for this reason, the nucleic acids form the genetic material of living cells which is passed on from generation to generation as the cell reproduces, so that the offspring have the ability to make the same proteins as their parents.

Discovery of the nucleic acids

Because it is such a fascinating story of scientific investigation, we shall, in this chapter, trace the discoveries and researches which have led up to our present day knowledge of the structures and the functions of the nucleic acids. It should be remembered that, in common with all lines of research, the elucidation of the structures of the nucleic acids was a long, difficult and sometimes frustrating task. Many, many scientists each made their small contributions, some correct, some incorrect; and, although negative results are often as valid in scientific investigation as positive ones, we have only space here to consider the work and advances of a few of the major investigators.

Nucleic acids were first discovered in 1868 by Friedrich Meischer, a Swiss medically trained chemist. After he had finished his medical studies he decided to investigate the chemistry of living cells, instead of going into clinical medicine as his father wished. Using pus from discarded hospital bandages as his research material, he managed to isolate a compound from the nuclei of the pus cells, and he called this compound nuclein. When he analysed this nuclein he found that it contained the elements carbon, hydrogen, oxygen, nitrogen and phosphorus. At that time the only other biochemical which was known to contain phosphorus was the compound lipid lecithin, but Meischer found that the proportion of phosphorus to nitrogen in his nuclein was very different from the proportions of these elements in lecithin. Quite clearly he had found a new biological compound.

Later, Meischer used the sperm of Rhine salmon as his research material, because he had observed that these cells were composed almost entirely of nuclei, and it was from the nuclei of his pus cells that he had originally isolated nuclein. He was able to confirm that nuclein was not lecithin, as it contained too much phosphorus. He also found that the phosphorus was present in nuclein as phosphoric acid. When he turned to the sperm of other animals, carp, frogs and bulls, he arrived at the same results. But it was left to another scientist, Altmann, to discover that nuclein had acidic properties, and to suggest in 1889 that the compound be renamed nucleic acid. Meischer's other major contribution, largely ignored at the time, was to suggest that the genetically active material supposed to be present in the nuclei of sperm cells was in fact nuclein.

Even before Meischer died in 1895, other scientists had begun to investigate the chemical nature of nucleic acid. In 1874, Piccard found a nitrogenous base in nucleic acid isolated from salmon sperm. This base was a purine compound, a double ring structure consisting of a six-membered heterocyclic ring of four carbon atoms and two nitrogen atoms, joined to a heterocyclic ring containing two more nitrogen atoms, and the compound which Piccard isolated was called guanine. Different purine bases are recognized by their having different atoms or side chains attached to the basic ring structure.

Then in 1880, a German chemist, Emil Fischer, demonstrated that there were two types of nitrogenous base present in nucleic acid, purines and pyrimidines. Pyrimidines are also heterocyclic ring compounds, but composed of just one ring with four carbon atoms and two nitrogen atoms, similar in fact to the six-membered ring part of the purine. Some years later Kossel isolated another purine base, called adenine, and two different pyrimidine bases called thymine and cytosine. And at the beginning of the twentieth century, Ascoli discovered yet another pyrimidine compound called uracil. Five major nitrogenous bases had now been identified from nucleic acid, and we now know that it is these five which make up the greater proportion of the nitrogenous base parts of the nucleic acids. In some cases other bases are present, but only occasionally and in small quantities.

Nitrogenous bases and phosphoric acid were thus known to be present in nucleic acid, but nothing was known of how these compounds were joined together in the acid. Nor did these components even represent the whole composition of the nucleic acids, for in 1910, a Russian born chemist, Levene, isolated a sugar molecule from nucleic acid extracted from yeast cells, a pentose sugar called ribose. Later he found that there were two types of nucleic acid, one containing the ribose sugar, and another one containing another pentose sugar called deoxyribose. Levene added a further contribution by suggesting a structure for the nucleic acid molecules, that the sugar and the nitrogenous bases were joined together as complexes, and that these complexes were joined together by the phosphoric acid components, to form long chains. These complexes were given names; a sugar attached to a nitrogenous base was called a nucleoside, and a nucleoside joined to a phosphate group was called a

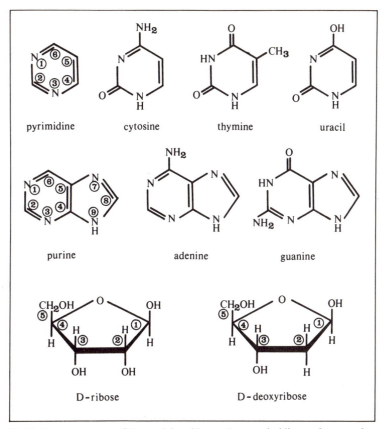

28. Major components of the nucleic acids are three pyrimidine and two purine bases, and two pentose sugars. DNA is composed of D-deoxyribose, with the bases adenine, guanine, cytosine and thymine, while RNA contains the sugar ribose instead of deoxyribose, and the base uracil in place of thymine

nucleotide. It is the nucleotides which form the very building blocks of the nucleic acid when they are joined together in long chains.

By 1930 it was clear that there were two distinct types of nucleic acid, which differed not only in the type of sugar present, but also in the composition of the nitrogenous bases. One of the types was originally found only in plant material, and was hence called plant nucleic acid. It was found to contain the sugar ribose, phosphoric acid, the two purine bases adenine and guanine, and the two pyrimidine bases cytosine and uracil. The other type of nucleic acid was isolated first from animal material and for a long time found only in animal cells, and was thus called animal nucleic acid. In this type, there was no ribose, only the pentose sugar deoxyribose, and while three of the bases, adenine, guanine and cytosine, were identical with those found in the plant nucleic acid, the uracil was replaced by another pyrimidine, thymine.

The distinction between plant and animal nucleic acids was soon proved to be inadequate, as every cell, plant or animal, was found to contain both types. So the names were changed to refer to the identity of the sugar group present. Plant nucleic acid thus became named ribose nucleic acid, later shortened to ribonucleic acid, and usually known today by the simple abbreviation RNA. Similarly animal nucleic acid was renamed deoxyribose nucleic acid, shortened to deoxyribonucleic acid, and referred to by its abbreviation DNA. And in the 1940s it was found that the DNA of a cell was localized within the nucleus, which explains why this is the compound which Meischer first isolated and named nuclein, while the RNA was found to be localized mainly in the surrounding cytoplasm.

The nucleotides

The general structure of the building blocks of the nucleic acids is the same for both RNA and DNA, and is shown in diagram 28. Nucleosides are composed of a sugar molecule, present in the five-membered furanose ring form, joined to one of the nitrogenous bases, the base being attached to carbon atom number one in the furanose ring of the sugar. Purines are always attached to the sugar through the nitrogen atom at position number nine in their rings, while the pyrimidines are also attached through a nitrogen atom, this time at position number three. Purine and pyrimidine bases do not usually occur free in nature, and are nearly always attached to a pentose molecule in this way.

Nucleosides are named according to the base involved as well as according to the sugar. The ribose nucleosides are named adenosine, guanosine, cytidine and uridine, and the names of the deoxyribose nucleosides are derived simply by adding the prefix deoxy- to the name of the corresponding ribose compound. Hence nucleosides formed with deoxyribose are called deoxyadenosine, deoxyguanosine and deoxycytidine, but because the pyrimidine thymine only forms a nucleoside with deoxyribose in nature, this nucleoside is called simply thymidine.

The nucleotides can be thought of as phosphate esters of the nucleosides. In the ribose ring there are three possible sites where a phosphate group could be attached, at carbon atoms numbers two, three and five. But in nature, only positions three and five are actually involved in these linkages. Nucleosides are named according to the nucleoside present and the position of the phosphate group, so that an adenosine nucleoside with a phosphate attached at carbon atom number three on the furanose ring is called adenosine-3-phosphate and cytidine with a phosphate on carbon atom number five would be cytidine-5-phosphate, and so on. In the deoxyribose nucleoside of course there is only a hydrogen atom, not a hydroxyl group, joined to the furanose ring at carbon atom number two, so only the other two carbon atoms, three and five, are able to join to phosphate groups anyway.

A few nucleotides have important functions of their own in living systems, apart from being part of the nucleic acids. In these cases, other phosphate groups are attached to the already present phosphate group on carbon atom number five, forming nucleotide-5-diphosphates if two phosphate groups are attached, and nucleotide-5-triphosphates if three phosphates are involved. Uridine-5-diphosphate for example plays a vital role in the metabolism of the carbohydrates. Joined to monosaccharide molecules, they form uridine diphosphate sugars (often abbreviated to UDP-sugars), and sugars joined in this way are more able to undergo the many various reactions which make up metabolism. In fact the nucleotides are acting in this way as supplements to the enzymes, for they have the effect of activating the sugars to undergo biochemical reactions, and so they are called co-enzymes.

The most important of the nucleotide phosphates is adenosine-5-triphosphate, usually abbreviated as ATP, for this molecule is able to store a large amount of energy and to release it on demand to the chemical machinery of the cell, to be used to provide energy for the biochemical reactions. It is a common feature of chemistry, organic chemistry and biochemistry that for many reactions to take place at all, they must be provided with an external source of energy, even in the presence of catalysts or enzymes. In the laboratory it is a simple matter to supply energy, in the form of heat energy, to a mixture of compounds so that they undergo a reaction, simply by placing a bunsen burner underneath the flask. A living cell, however, is not able to apply a bunsen burner to itself, and so it must look elsewhere for its energy. This is supplied in the form of chemical energy from food materials such as carbohydrates and fats; but because the living cell does not necessarily wish to use all the available energy at one time, it must be stored for some future time. There is little point in storing a large amount of energy in one 'package', which, when released, must either be used all at once, wasted, or re-stored. So the cell stores energy in many small 'packages'; then, when it requires some energy for a particular reaction, it can release just enough from one or two 'packages'. Thus energy is not wasted, nor does it need to be re-stored, a process which itself uses up a little energy.

Energy is stored in the living cell in the chemical bond between the second and the third phosphate groups in the ATP molecule. When the cell requires some energy, it breaks this compound down to adenosine diphosphate (ADP), thus releasing a phosphate group and the discrete amount of energy stored in the bond. The chemical bond between the two phosphate groups in ATP is sometimes, not quite correctly, called a 'high energy bond', and the ATP itself has been called the 'universal cell fuel'. We can see the effects of ATP in action in the luminous tails of the firefly insects, which obtain the energy necessary for the luminous effect from the breakdown of ATP to ADP. When the cell wishes to store energy, perhaps from the food materials it has broken down, it carries out a reverse of the above process, joining a phosphate group on to a molecule of ADP, and storing the energy in the bond between the phosphate groups at the same time.

Polynucleotides

When nucleotides join together to form long chains, both the possible hydroxyl groups in the sugar group are involved in phosphate ester formation. Thus the long polynucleotide chain consists of a 'backbone' of sugar groups joined together through phosphate groups which form bridges between carbon number three on one sugar and carbon number five on the next, and so on. And the bases project perpendicular to this backbone, attached to the sugar groups. In this way, tens, hundreds or thousands of nucleotides are joined together to form polynucleotide chains which are the basis of the

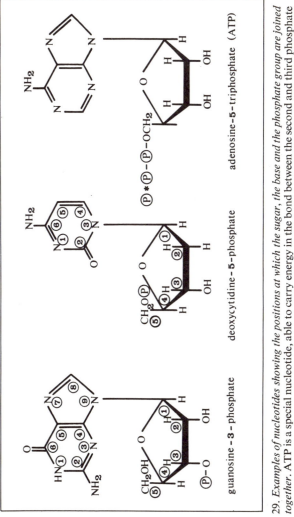

29. *Examples of nucleotides showing the positions at which the sugar, the base and the phosphate group are joined together. ATP is a special nucleotide, able to carry energy in the bond between the second and third phosphate groups (marked as a star). When this bond is broken to split off a phosphate group, adenosine-5- diphosphate (ADP) remains, and the energy in the bond is released for use in biochemical reactions*

RNA and the DNA molecules. This much of the structure of these molecules was known in the early part of the twentieth century, but it was not known how many nucleotides were involved in a molecule of a nucleic acid, what factors determined the sequence of the different bases, or nucleotides, along the chain—the primary structure—nor was the shape, or secondary structure, of the molecule known. Because the bases were found to be present in roughly equal amounts, it was suggested that four different nucleotides were joined together in a tetranucleotide, but as methods of analysis advanced, it was found that all the bases were not in fact present in equal amounts, and that the molecular weights of the molecules were much higher than would be expected for a tetranucleotide. This was not the answer.

Structure of DNA

As methods for the extraction of DNA improved so that this material could be extracted without damage, estimates for its molecular weight increased steadily, and by the early 1950s it was clear that these molecules were very large indeed, having molecular weights of the order of ten million, so that they are composed of some 50,000 nucleotide units joined together. But in many ways the most important piece of research which led up to the discovery of the structure of these molecules was that carried out by Edwin Chargaff in New York about 1950. Chargaff and his colleagues analysed the base composition of DNA which they extracted from many sources, and from their results devised a set of 'rules' for the structure of the molecules. They found that the base composition of DNA is characteristic for a particular species, and that different cells or tissues of any species had identical, or at least very similar, base compositions in the DNA. But in all samples of DNA, from all sources, they found that there were a number of regularities. The number of purine bases always equalled the number of pyrimidine bases, and furthermore the number of adenine bases equalled the number of thymine bases, and the number of cytosine bases equalled the number of guanine bases. Why these regularities occured, Chargaff could not discover, but without these 'rules' the final discovery of the structure of the DNA molecule would have been a much more difficult task.

As well as chemical regularities, physical regularities were also found in DNA molecules. As these compounds are acids, they can form salts, and the sodium salts of DNA can be spun into long fibres. And because such fibres have definite crystal structures, it was only natural that scientists should begin to look at them with X-ray diffraction methods, similar to those which had been applied to the problem of the structure of proteins. One scientist, Maurice Wilkins, working in London, found that these fibres exhibited a repeating structural regularity every 34 Å (one Angstrom unit [Å] is 10^{-10} metre), and another regularity every 3.4 Å along these fibres. He postulated that the nucleotide chain was twisted in some way, and that the twists accounted for these structural regularities. Wilkins also performed some density measurements on DNA molecules, and the results from these indicated to him that each molecule of DNA consists of more than one single polynucleotide chain.

The stage was now set for some scientist to come along, look at all the evidence which had been collected together, and suggest a structure for DNA. Pauling and Corey had, at about this time, suggested their helix structure for proteins, and they therefore put forward a similar suggestion for DNA. But in their structure they envisaged three polynucleotide chains twisted together in a helix, and this structure did not fit with Chargaff's rules nor with the X-ray results. Then in 1953 James Watson and Francis Crick, working at Cambridge University, put forward their by now famous double helix structure for DNA, a structure which fitted all the evidence and which also explained many of the suspected functions of the nucleic acids. For this work they, together with Wilkins, were awarded the Nobel prize in 1962, and although there have been some modifications, their structure is still generally accepted today.

Watson and Crick's double helix model looks rather like a ladder which has been twisted about a central axis. The two 'uprights' of the ladder are formed by the backbones of sugar and phosphate groups, while the bases are turned inwards from these backbones to form the 'rungs'. A base on one polynucleotide chain is joined to a base on the other chain by hydrogen bonds, so that the two halves of the rungs are stuck together in a stable structure. In this structure each of the 'rungs', or base pairs, is 3.4 Å from the next one, and each turn of the helix has a height of 34 Å, corresponding to one turn of the helix every 10 nucleotides, thus explaining the results from the X-ray diffraction studies.

The joining of the bases, the purines and the pyrimidines, in the centres of the base pairs is very specific. For all the groups, sugars, phosphates and bases, to fit neatly into the helix, an adenine base can only form a base pair with a thymine base, and a guanine can only pair with a cytosine. If the two polynucleotide chains were arranged so that, say, an adenine lay opposite a cytosine, these two bases

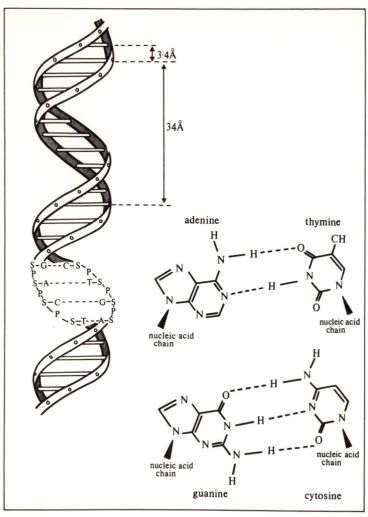

30. *Double helix structure of the DNA molecule*, showing the dimensions and arrangements of the sugars, the phosphates and the bases. The bases are joined together in the helix by hydrogen bonds as shown in the structures at right

could not form hydrogen bonds between themselves. If two purines, the larger of the bases, lay opposite each other, there would not be sufficient room to accommodate them both in the double helix structure. In fact there are valid objections to any base pairs being formed, except those mentioned above. So because an adenine always pairs with a thymine, there must be equal numbers of each of these bases in the whole molecule, which agrees with the result obtained by Chargaff. A similar situation of course occurs with guanine and cytosine; because one is always paired with the other, there must be an equal number of each present. Therefore each chain in the double helix is said to be complementary to the other; one chain with a sequence guanine — cytosine — adenine — guanine — thymine will always lie opposite and be paired by hydrogen bonds with a sequence cystosine—guanine—thymine—cytosine—adenine, and this same specific pairing occurs all the way along the molecule. One further implication of this structure is that for hydrogen bonding to occur, one of the bases must be 'upside down' in relation to the other, so that in fact the whole chain is 'upside down' in relation to the other.

The other nucleic acid, RNA, differs structurally from DNA in a number of respects. First of all, the sugar is different, and instead of the pyrimidine thymine, there is uracil. But base pairing does not occur in this molecule, as it is only a single chain of nucleotides, only one half of the ladder. For this reason there is very little helical structure to the RNA molecule, and it is usually coiled or folded in some complex, irregular way.

Functions of DNA

It was mentioned at the beginning of the chapter that the nucleic acids contain in their structures all the instructions for the synthesis of the proteins, instructions for the way in which the amino acids must be arranged in a sequence and joined together. Meischer had been the first to suggest that his nuclein was a likely candidate for the genetic material of living cells, by then supposed to be localized in the nucleus of a cell. But his suggestion had been largely forgotten, and it was not until 1944 that there was an actual demonstration that DNA was responsible for carrying genetic information. In that year, O. T. Avery and his co-worker extracted the DNA from a strain of bacteria called pneumococcus which had fully developed capsules around the outside of their cells, and were therefore called 'smooth' cells. When they introduced this DNA into some pneumococcus cells which lacked capsules, they found that some of the offspring of these 'rough' cells were transformed into 'smooth' cells with capsules. Clearly it was the DNA which had enabled these 'rough' cells to grow capsules, and which had carried the genetic information to the new offspring. But the actual mechanism of this inheritance could not be understood until after Watson and Crick had discovered the structure of the DNA.

Watson and Crick were aware that their structure for DNA suggested a method by which information could be carried and passed on from generation to generation, and they were careful to point this out when they published the details of the double helix. They suggested that in some way the sequence of the different bases along the DNA polynucleotide chain provided coded information for the arrangement of the amino acids in proteins. So if these coded instructions can be passed out to the protein synthetic machinery in the cytoplasm of the cell and translated there, specific proteins, with specific amino acids sequences, could be made according to the instructions. But also involved in this transmission and translation of information is the way in which the DNA molecule forms copies of itself, for it is the copies, and not the original DNA molecules, which carry the information, and which are also distributed to the next generation of cells when the original cell divides and reproduces.

If the two strands of the DNA double helix can be made to unwind, and if there is a ready supply of each of the four different types of nucleotides, they could line up along the unwound DNA strand according to Chargaff's rules for base pairing. If these nucleotides can then join together, and also form hydrogen bonds with their opposite bases, two new strands of double helix DNA could be made, each new molecule containing one strand from the original molecule and one new strand, and each complete DNA double helix so produced being identical to the original. The DNA strand which was part of the original molecule, and on which the nucleotides line up and join together, is called the 'template' strand, because it acts as the model for the arrangement of the bases in the complementary strand. The double DNA molecule can also act as the model for the synthesis of an RNA molecule. In this case the DNA molecule unwinds as before, but only one strand acts as a template for the production of a single strand of RNA, and instead of thymine nucleotides lining up opposite adenine bases in the template, uracil nucleotides are involved. Both of these processes occur in living cells. DNA makes a copy of itself which is passed on to the next generation of cells as it divides. And by making RNA, there is a method by which the coded instructions on the DNA molecule can be passed out into the cytoplasm of the cell. It will be remembered that DNA was found to be localized mainly in the nucleus of cells, while RNA was present in greater proportions in the surrounding cytoplasm.

The genetic code

The only problems now remaining are how the DNA double helix carries the genetic information in its structure and how this information is translated in the protein synthetic machinery of the cytoplasm. Watson and Crick suggested that the information was carried in the sequence of the bases along the DNA polynucleotide chain. But there are only four different bases, and these must somehow provide an unambiguous specification for some twenty different amino acids. If the arrangement is that a single base in the DNA molecule provides a specification for one amino acid in a protein chain, that is, if the amino acids line up along an RNA strand in the cytoplasm, one opposite just one base, then only four different amino acids can be specified by the nucleic acid structure. Clearly this is not enough. If a combination of two bases specify one amino acid, the situation improves. Four different bases can be arranged in pairs in 16 different ways (4 × 4), for example AG,AC,GA,CA,AU,AA,UA and so on. If in the protein synthetic machinery in the cytoplasm, the amino acids line up along a strand of nucleic acid such that each amino acid is specified by a pair of bases, then the nucleic acid can provide specifications for sixteen amino acids. But there are still about four amino acids which cannot be specified according to this scheme, so this is not the answer either.

The third possibility is that a series of three bases in the nucleic acid structure specifies each amino acid. Four bases can be arranged in triplets in 64 different ways (4 × 4 × 4), and so there are more than enough possibilities to specify for the amino acids in proteins. Therefore, in the protein synthetic machinery in the cytoplasm, each amino acid is lined up along the nucleic acid chain opposite a series of three bases, so that when all the amino acids are joined together,

THE GENETIC CODE

Second Base

		U		A		G		C		
		UUU	Phe	UAU	Tyr	UGU	Cys	UCU		U
	U	UUA ⎫	Leu	UAA ⎫	Stop	UGA	Stop	UCA	Ser	A
		UUG ⎭		UAG ⎭		UGG	Try	UCG		G
		UUC	Phe	UAC	Tyr	UGC	Cys	UCC		C
		AUU ⎫		AAU	Asn	AGU	Ser	ACU		U
	A	AUA ⎭	Ile	AAA ⎫	Lys	AGA ⎫	Ary	ACA	Thr	A
		AUG	Met	AAG ⎭		AGG ⎭		ACG		G
		AUC	Ile	AAC	Asn	AGC	Ser	ACC		C
First base		GUU		GAU	Asp	GGU		GCU		U
	G	GUA	Val	GAA ⎫	Glu	GGA	Gly	GCA	Ala	A
		GUG		GAG ⎭		GGG		GCG		G
		GUC		GAC	Asp	GGC		GCC		C
		CUU		CAU	His	CGU		CCU		U
	C	CUA	Leu	CAA ⎫	Gln	CGA	Arg	CCA	Pro	A
		CUG		CAG ⎭		CGG		CCG		G
		CUC		CAC	His	CGC		CCC		C

Third base

31. *The Genetic Code*, the triplets of bases in the messenger RNA molecule which specify which amino acids are joined together in a protein chain. Only 61 of the possible 64 triplets actually specify amino acids, and the remaining three, marked as 'stop', tell the protein synthetic machinery when it has come to the end of a particular protein chain. The abbreviations for the amino acids are the same as those used in Chapter 5

they will form a specific protein according to the arrangements of the base triplets along the nucleic acid molecule. This arrangement is called the genetic code, or the triplet code, and has been confirmed by complex and intricate experiments. All the triplets, together with the amino acids which they specify, are shown in diagram 31. It will be noticed that three of the triplets are not codes for amino acids, but are labelled 'stop', and these triplets are situated at the beginning or the end of a sequence of triplets, telling the protein synthetic machinery in the cytoplasm when to begin and end the synthesis of a particular protein. Were these triplets not present, long chains of amino acids would be produced, consisting of a number of proteins all joined end to end, and the cell would not know where to break the chain to obtain the individual proteins.

Protein synthesis

Finally we come to the method by which the coded information in the genetic code is actually translated in the cytoplasm to make proteins. The double helix DNA in the nucleus makes a number of different RNA molecules, each with its own specific function in protein synthesis. The first is called ribosomal RNA, and these molecules are destined to form small particles called the ribosomes, which will be the synthetic factories in which the proteins are made. Ribosomal RNA molecules are highly folded and coiled so that they are able to make the almost spherical ribosomes.

The second type of RNA is called messenger RNA. This is a straight molecule, with a base sequence which is complementary to a particular section of the DNA molecule which contains the information for a particular protein, and, as the name suggests, these molecules are responsible for taking the coded instructions from the nucleus out into the cytoplasm and the ribosomes, where they will be translated. The third type is called transfer RNA. This is also a folded and coiled molecule, but there are two special regions of the molecule which have definite functions. One of these regions is just a series of three bases which correspond to a triplet which specifies a particular amino acid. The other region of the transfer RNA is able to attract, and hold, the amino acid which corresponds to, and is specified by, the triplet at the other region. It is these transfer RNA molecules which are ultimately responsible for translation of the genetic coded message. With their amino acids on their backs, they line up along the strand of messenger RNA so that the bases in the triplet region can form pairs according to Chargaff's rules with a complementary section of the messages. Thus a triplet on the messenger RNA holds a transfer RNA with a complementary triplet, which in turn holds the amino acid specified by that triplet. Another transfer RNA, with a different triplet and hence a different amino acid on its back, can then line up on the next triplet on the messenger, and the two amino acids can then join together. When the first transfer RNA has been released by the messenger RNA, another transfer RNA with another amino acid can line up and add its amino acid to the other two. All this happens inside the ribosome, so that as the ribosome moves along the messenger RNA strand, transfer RNA molecules come in, form transient base pairs with their complementary triplets on the messenger strand, give up their amino acids to the growing amino acid chain, and are then released. By the time the ribosome has travelled the length of the messenger RNA molecule, or until it reaches a triplet which spells stop, it will have collected all the amino acids for a particular protein, all in the correct order, and joined them together. The protein can then be released to coil up into its characteristic secondary and tertiary structures, and the ribosome is freed to find another messenger RNA molecule to begin the process over again and make another protein.

Summary

We have now looked at all the major classes of biochemical compounds which make up living organisms, and the ways they are inter-related to give the properties we call life. The nucleic acids contain all the information which living cells need to be able to make their structures, for they control the production of the proteins, which either form structures themselves, or act as enzymes which control the biochemical reactions in which other materials are broken down or built up into living structures. The carbohydrates and the lipids have functions as structural materials or as stores of food materials, and hence, ultimately, as stores of energy, and all the reactions and transformations of these chemicals are under the control of the enzymes. All these biochemical reactions are inter-related and are called collectively the metabolism of living organisms, but they must form the subject of another book.

Index